WAR AMONGST THE PEOPLE

SANDHURST TRENDS IN INTERNATIONAL CONFLICT

Edited by David Brown, Donette Murray, Malte Riemann, Norma Rossi and Martin A. Smith, The Royal Military Academy Sandhurst, UK

The Sandhurst Trends in International Conflict series is a cutting-edge forum and platform for original thought and debate on military and security matters within the contemporary international security environment. It aims to stimulate authors to think critically about contemporary conflict and security more generally and to identify and evaluate practical, political and doctrinal lessons from recent experience. The Sandhurst series invites practitioners and academics from the fields of security, diplomacy, the law, politics and the military to interrogate and publish on the key debates that will shape both the contemporary international security environment and a modern military operating within it.

War Amongst the People

Critical Assessments

Edited by

DAVID BROWN

DONETTE MURRAY

MALTE RIEMANN

NORMA ROSSI

MARTIN A. SMITH

Howgate Publishing Limited

First published in 2019 by
Howgate Publishing Limited
Station House
50 North Street
Havant
Hampshire
PO9 1QU
Email: info@howgatepublishing.com
Web: www.howgatepublishing.com

British Library Cataloguing-in-Publication Data
A catalogue record for this book is available from the British Library

ISBN 978-1-912440-02-3 (pbk)
ISBN 978-1-912440-03-0 (ebk - PDF)
ISBN 978-1-912440-18-4 (ebk - ePUB)

The views expressed in this book are those of the individual authors and do not necessarily reflect official policy or position.

CONTENTS

FIGURES AND TABLES

Figures

Tables

ABOUT THE EDITORS

Dr David Brown is a Senior Lecturer in the Department of Defence and International Affairs at the Royal Military Academy Sandhurst. He has written extensively on a range of security related issues, publishing books and articles on US and UK foreign and defence policy, contemporary power relations, aspects of European security and international intervention.

Dr Donette Murray is a Senior Lecturer at the Royal Military Academy Sandhurst. A Fellow of the Supreme Headquarters Allied Powers Europe (SHAPE) and a former political advisor, she holds a doctorate from the University of Ulster and an LLM in International Law from the University of Maastricht. The author of four books on US foreign policy, she has also published in the *Hague Yearbook of International Law* on states' use of force in self-defence against non-state actors.

Dr Malte Riemann is a Senior Lecturer in the Department of Defence and International Affairs at the Royal Military Academy Sandhurst. Dr Riemann studied in Bremen and Pietermaritzburg and holds a PhD in International Relations from the University of Reading. His fields of interest include the privatisation of war and its effects on the state's legitimate monopoly on violence, the medicalisation of security, and the historicity of non-state actors.

Dr Norma Rossi is a Senior Lecturer in Defence and International Affairs at the Royal Military Academy Sandhurst. Dr Rossi studied in Rome and Paris and received her PhD in Politics and International Relations from the University of Reading on an Earhart Foundation fellowship. Recently, she has extended her research interests to include Security Sector Reform and the role of education in conflict-affected environments.

Dr Martin A. Smith is Senior Lecturer in Defence and International Affairs at the Royal Military Academy Sandhurst. Prior to joining RMAS he was at the Department of Peace Studies University of Bradford, from where he received his PhD in 1994. His main research interests are in the fields of international power, European security and US foreign policy.

CONTRIBUTORS

Major John Bailey is a British infantry officer with operational experience in Iraq, Afghanistan, Northern Ireland and Cyprus. He has an MA in History (2001) and an MLitt in International Relations (2013). Both were completed at St Andrews University. He is now a PhD candidate at Exeter University. In his most recent military staff appointment he was on the writing team for Army Field Manual, *Tactics for Stability Operations*.

Lt Col. Grant Davies is a serving Army Legal Services officer with 20 years' experience, having served in Northern Ireland, Kosovo and Afghanistan. He has fulfilled a number of roles, including the senior legal advisor in Regional Command (South) in Kandahar, Afghanistan, and team leader of the Detention Oversight Team, Afghanistan. A solicitor advocate, he holds a degree in International Relations and Strategic Studies from UCW Aberystwyth, and a Masters Degree in Public International Law from King's College London.

Whitney Grespin was awarded a BA by Duquesne University and a Masters in Public and International Affairs by the University of Pittsburgh. She is currently a PhD student at King's College London's Defence Studies Department and teaches at the United Kingdom Defence Academy's Joint Services Command and Staff College. Her academic work focuses on the U.S. Government's use of contractors to deliver foreign military training. She also works as Peace Operations Analyst for the U.S. Army War College's Peacekeeping and Stability Operations Institute (PKSOI) and is a Term Member at the Council on Foreign Relations.

Beatrice Heuser is Professor of International Relations at the University of Glasgow. She holds degrees from the Universities of London (BA, MA) and Oxford (DPhil), and a Habilitation from the Philipps-University of Marburg. Previously, she has taught at the Department of War Studies, King's College London, the University of Reading, Sciences Po, the Sorbonne, and in Germany. She has also worked on the International Staff at NATO headquarters in Brussels and as Director for Research at the Military History Research Office of the Bundeswehr in Potsdam. Professor Heuser is the author of numerous publications, including: *NATO, Britain, France and the FRG: Nuclear Strategies and Forces for Europe* (1997); *Nuclear Mentalities? Strategies and Belief-Systems in Britain, France and the FRG* (1998); *The Bomb: Nuclear Weapons in their Historical, Strategic and Ethical Context* (1999); *Reading Clausewitz* (2002); *The Evolution of Strategy: Thinking War from Antiquity to the Present* (2010); and *Strategy before Clausewitz: Linking Warfare and Statecraft 1400–1830* (2017).

Georgina Holmes is a Leverhulme Early Career Research Fellow in the Department of Politics and International Relations at the University of Reading. Her research examines the development of gender mainstreaming as a policy frame in global governance and peacekeeping, norm implementation dynamics, and the training and deployment of African and European uniformed personnel in peacekeeping operations. Dr Holmes has published articles in several peer reviewed academic journals including *International Peacekeeping*, *Journal of Intervention and Statebuilding*, *Journal of Genocide Studies* and *The RUSI Journal* and is the author of *Women and War in Rwanda: Gender, Media and the Representation of Genocide* (2013).

Andree-Anne Melancon is a Senior Lecturer at the Royal Military Academy Sandhurst. She holds a PhD in Politics from the University of Sheffield and her thesis was titled 'Discrimination in the Age of Drone Warfare'. Dr Melancon's main research interest focuses on military ethics, the just war tradition, and drone warfare.

Lt Col. Jyri Raitasalo PhD, is Military Professor of War Studies at the Finnish National Defence University (FNDU) where he holds the title of Docent of Strategy and Security Policy. Among his recent appointments, Jyri has served as senior staff officer (strategic planning) at the Finnish MOD, the Commanding Officer of the Helsinki Air Defence Regiment (Armoured Brigade), Head Lecturer of Strategy at the Finnish National Defence University, and ADC to the Chief of Defence and Staff Officer (strategic planning) in the Finnish Defence Command.

Vladimir Rauta is a Lecturer in Politics and International Relations at the University of Reading. His research focuses on proxy wars and external support to violent parties in civil wars. Dr Rauta is currently researching the micro dynamics of political violence in proxy wars with a focus on Africa and the Middle East. His research is forthcoming in journals such as *International Relations* and *Studies in Conflict and Terrorism*.

Malte Riemann is a Senior Lecturer in the Department of Defence and International Affairs at the Royal Military Academy Sandhurst. Dr Riemann studied in Bremen and Pietermaritzburg, and holds a PhD in International Relations from the University of Reading. His research has been funded by the German Academic Exchange Service, the Earhart Foundation and the European International Studies Association. His fields of interest include the privatisation of war and its effects on the state's legitimate monopoly on violence, the medicalisation of security, and the historicity of non-state actors. He is currently in the process of writing a monograph in German on the transformation of war provisionally titled 'War in the 20th and 21st Century', and his most recent publication 'Problematizing the medicalization of violence: A critical discourse analysis of the "Cure Violence" initiative' has just appeared in *Critical Public Health*.

Norma Rossi has been Senior Lecturer in Defence and International Affairs at the Royal Military Academy Sandhurst since 2014. Dr Rossi studied in Rome and Paris, and received her PhD in Politics and International Relations from the University of Reading on an Earhart Foundation Fellowship. She works on state-building processes and has specific expertise on the role of transnational non-state actors, and the link between organised crime groups and state-building. She has also written on the politics of anxiety with particular attention to the relation between the far right and neoliberal politics in Italy. Since completing her PhD, she has broadened her research interests to include security sector reform and the role of education in conflict-affected environments. She has co-designed and delivered the International Conflict Management course in Croatia, Hungary, Lebanon, Chile and the Occupied Territories of Palestine.

Alex Waterman holds a PhD from the University of Leeds and is a contributor to the International Institute for Strategic Studies Armed Conflict Database and Associate Reviews Editor for the journal *Civil Wars*. His research focuses on the relationship between counterinsurgency and state formation, state-building and legitimation in complex political orders. His doctoral thesis explores counterinsurgents' attempts to shape and negotiate order in Northeast India. Dr Waterman is also interested in utilising the concept of order to generate insights into rebel governance, intra-state conflict, terrorism and hybrid warfare in South Asia and further afield.

Major Ian Wilson is a serving British Army officer in the Royal Regiment of Scotland with 19 years' military leadership. He has first-hand experience of Britain's foreign policy over the last two decades, including deployments to Bosnia, Iraq and Afghanistan. Major Wilson holds Masters degrees from the University of Exeter and King's College London. He was awarded the 2012 Fuller Prize for his academic work in Defence Studies while on the Intermediate Command and Staff Course. His research interests include British foreign policy, democracies at war and global political power relationships.

FOREWORD

Writing a Foreword to this book, the first in the Sandhurst Trends in International Conflict series, gives me great pleasure.

In the first instance, the idea, albeit not new, of such a publication to provide for wider debate is a good one. It is particularly welcome at this time, when the British Army – small, under resourced and facing a wide range of potential threats – must think hard and deeply if it is to be successful in the future.

What better place to start doing this than at Sandhurst? Not only must this searching inquiry take place, but its outcome must also be put into effect. The Army is now too small to allow for separate elements, each formed for a particular style of war, for example Armoured or Light, Conventional or Counter Insurgency. The whole army must be able to operate to one purpose against whatever foe, wherever they are located. To assume otherwise is to plan by design that we will not concentrate our force; for a small army this leads to defeat.

To be able to concentrate the force, as opposed to massing the forces, will require considerable changes to organisations, methods of command and control, tactics, weapon requirements and so on. Making these changes successfully will require leadership. It will require leaders who comprehend the rationale for change and the underpinning concepts of the change.

Secondly, it is pleasing that the idea of 'war amongst the people', which I introduced in 2005 with the publication of my book, *Utility of Force*, is still of continued contemporary relevance.

The phrase 'war amongst the people' was coined by me to describe a form of war, not battle. War has a political aim borne in a confrontation over some issue and, having failed to resolve the confrontation by pacific means, armed force is employed at least in part to gain the desired political outcome. Battles, on the other hand, have military objectives. They can be of any size, although the trend is to employ relatively small groupings, even if large numbers are deployed. Battles remain as bloody, as frightening and as exciting as they have ever been.

These wars take place 'amongst the people' in a number of ways, each war presenting a different combination of the following elements. Wars are being fought to have control over or the loyalty of the people, rather than having direct control of territory. To counter those who possess modern weapons and the ability to operate in the air and space, it is safer to conceal oneself in the 'clutter' of the people in urban areas. Lastly, there is the 'theatre' of war, where one seeks to convince the people, both those in and without the operational area, by word and deed, that you are the authority to follow.

One of the characteristics of 'war amongst the people' is that the use of armed force does not lead directly to a resolution of the underpinning confrontation – which is why every battle can be won and yet the war is lost. Our institutions, such as those of government and the armed forces, all developed in the context of Industrial War, the form of war that preceded 'wars amongst the people'. The new circumstances require these institutions to change.

I commend the papers herein for each in its way exposes discomfort with what has happened or is expected to happen. In a volume such as this discomfort is good, as it generates debate and leads to change.

General Sir Rupert Smith KCB DSO OBE QGM

ACKNOWLEDGEMENTS

Edited volumes are, necessarily, the product of a great deal of collaboration. *War Amongst the People* – the first book in the Sandhurst Trends in International Conflict series – has benefited from the support and expertise of many.

To begin with, we wish to thank the Royal Military Academy Sandhurst. As host and key proponent, RMAS has been instrumental in the success of the series.

We would like to thank our authors for their enthusiastic embrace of the multidiscipline approach and ethos we have sought to engender in this project.

Thanks are also due to those colleagues – old and new – from the military, government, think tanks, academia and elsewhere who have attended our biannual symposia. Lively, provocative and illuminating, these exchanges have helped to lay foundations for critical engagement and cooperation. They have also been very enjoyable!

Finally, we would like to take this opportunity to thank our publisher, Kirstin Howgate, for her creative input, unfailing patience and practical assistance.

DB
DM
MR
NR
MAS

January 2019

GLOSSARY

AU	African Union
APs	Additional Protocols (to the Geneva Conventions)
CA	Common Article (Geneva Conventions)
C4ISR	Command, Control, Communications, Computers, Intelligence, Surveillance and Reconnaissance
CIHL	Customary International Humanitarian Law
CIL	Customary International Law
COIN	Counterinsurgency
CPTMs	Core Pre-deployment Training Materials
CRSV	Conflict-Related Sexual Violence
DFS	Department of Field Support (UN)
DoD	Department of Defense (US)
DOTMLPF	Doctrine, Organization, Training, Manpower, Leadership and Education, Personnel, Facilities, and Policy
DPH	Direct Participation in Hostilities
DPKO	Department of Peace-Keeping Operations (UN)
ECHR	European Convention on Human Rights
ECtHR	European Court of Human Rights
FID	Foreign Internal Defense
FMF	Foreign Military Financing
FPR	Foreign Policy Restraint
GC	Geneva Conventions
GWOT	Global War on Terror
HI	Humanitarian Intervention
IAC	International Armed Conflict
ICCPR	International Covenant on Civil and Political Rights
ICJ	International Court of Justice
ICRC	International Committee of the Red Cross
ICTY	International Criminal Tribunal for the Former Yugoslavia

IDPs	Internally Displaced Persons
IED	Improvised Explosive Device
IHL	International Humanitarian Law
IHRL	International Human Rights Law
IMET	International Military Education and Training (US)
IPC	Internal Political Change
IRoS	Imperative Reasons of Security
IS	Islamic State (of Iraq and the Levant)
LOAC	Law of Armed Conflict
MoD	Ministry of Defence (UK)
NDAA	National Defense Authorization Act
NGOs	Non-Governmental Organisations
NIAC	Non-International Armed Conflict
PMCs	Private Military Companies
PoC	Protection of Civilians
POW	Prisoner of War
QIP	Quick Impact Project
RDF	Rwandan Defence Force
RMA	Revolution in Military Affairs
RoE	Rules of Engagement
R2P	Responsibility to Protect
SEA	Sexual Exploitation and Abuse
SFA	Security Force Assistance
SME	Subject Matter Expert
SOF	Special Operations Forces
SSR	Security Sector Reform
TAA	Train, Advise, Assist
TCCs	Troop Contributing Countries
UNAMID	United Nations – African Union Hybrid Operation in Darfur
UNSCR	United Nations Security Council Resolution
WPS	Women, Peace and Security

INTRODUCTION

'War Amongst the People': A Prism for Analysis?

Malte Riemann and Norma Rossi

[W]ar as a battle in a field between men and machinery, war as a massive deciding event in a dispute in international affairs: such war no longer exists.[1]

General Sir Rupert Smith

Since the end of the Cold War, debates on the changing character of war intensified amongst both academics and practitioners. There has been an emergence of diverse bodies of literature that, despite their (strong) differences in focus – particularly regarding the nature and implications of change – share the conviction that modern developments have led to fundamental changes in the nature of warfare. For some, war's character was changing due to a Revolution in Military Affairs (RMA) 'that transformed advanced state militaries through an emphasis on stripped-down, highly specialized forces deploying cutting-edge technology with unprecedented precision'.[2] While proponents of the RMA thesis see material factors determining the conditions of change in the character of war,[3] another view sees its transformation more in relation to the actors

---

1 Rupert Smith, *The Utility of Force: The Art of War in the Modern World* (London: Penguin, 2006), 1.

2 Derek Gregory, 'War and Peace', *Transactions of the Institute of British Geographers* 35:2 (2010), 1.

3 Jean Baudrillard, *The Gulf War Did Not Take Place* (Bloomington: Indiana University Press, 1995); Norman Davis, 'An Information-Based Revolution in Military Affairs', *Strategic Review* 24:1 (1996), 43–53; Michael Ignatieff, *Virtual War: Kosovo and Beyond* (London:

involved and the territories in which it is fought.[4] Mary Kaldor, one of the leading advocates of this position, argues that 'New wars are fought by varying combinations of networks of state and non-state actors ... New wars are fought in the name of identity ... [and] battles are rare and territory is captured through political means, through control of the population'.[5] In line with Kaldor, Steven Pinker claims that interstate conflict is in decline, heading towards obsolescence.[6] This thesis has resonated widely with Western militaries. Since the early 2000s, the US, UK and France in particular have increasingly focused on population-centric warfare and continue to view 'the people' as central in an age of hybrid war.[7] As Lt Gen. William Caldwell states:

> The future is not one of major battles and engagements fought by armies on battlefields devoid of population; instead, the course of conflict will be decided by forces operating among the people of the world. Here, the margin of victory will be measured in far different terms than the wars of our past. The allegiance, trust, and confidence of populations will be the final arbiters of success.[8]

A series of contributions, often led by so-called 'warrior-scholars' such as John Nagl, Robert Cassidy, David Kilcullen and Vincent Desportes, as well as academics strongly embedded with the military, such as Montgomery McFate, are shaping an ever-growing counterinsurgency (COIN) literature, while also informing army doctrine.[9] Here, the idea of

Vintage Books, 2001); James Der Derian, *Virtuous War: Mapping the Military-Industrial-Media-Entertainment Network* (Boulder, Colorado: Westview Press, 2001).

4   Mary Kaldor, *New and Old Wars: Organized Violence in a Global Era* (Cambridge: Polity Press, 1999); Herfried Münkler, *Die neuen Kriege* (Reinbek bei Hamburg: Rowohlt, 2002); Thomas X. Hammes, *The Sling and the Stone: On War in the 21st Century* (St. Paul: Zenith Press, 2006); Peter A. Kiss, *Winning Wars amongst the People: Case Studies of Asymmetric Conflicts* (Lincoln: University of Nebraska Press, 2014).

5   Mary Kaldor, 'In Defence of New Wars', *Stability: International Journal of Security and Development* 2:1 (2013), 1.

6   Steven Pinker, *The Better Angels of Our Nature: A History of Violence and Humanity* (London: Penguin, 2012).

7   Christopher S. Chivvis, *Understanding Russian 'Hybrid Warfare' and what we can do about it.* Testimony before the Committee on Armed Services on 22 March 2017 (Washington DC: United States House of Representatives, 2017) available at https://www.rand.org/pubs/testimonies/CT468.html accessed on 21 December 2018; Uwe Hartmann, *The Evolution of the Hybrid Threat, and Resilience as a Countermeasure* (Rome: NATO Defense College, 2017); Maria Mälksoo, 'Countering hybrid warfare as ontological security management: the emerging practices of the EU and NATO', *European Security* 27:3 (2018), 374–92.

8   William B. Caldwell and Steven M. Leonard, 'Field Manual 3-07, Stability Operations: Upshifting the Engine of Change', *Military Review* 88:6 (2008), 6.

9   David Galula, *Counterinsurgency Warfare: Theory and Practice* (Westport: Praeger Security International, 2006); Robert M. Cassidy, *Counterinsurgency and the Global War on Terror*

conquering 'the hearts and minds' of the local population has become the central focus. This has challenged the scope and utility of conventional military force, with Nagl claiming that '[c]onventional armies are not well suited to the demands of counterinsurgency'.[10] Indeed, according to Frank Hoffmann, these conflicts pose a 'planning dilemma for today's military planners, raising a putative choice between preparing for states with conventional capabilities or the more likely scenario of non-state actors employing asymmetric or irregular tactics'.[11]

Of particular importance within this debate is Rupert Smith's *The Utility of Force: The Art of War in the Modern World*. In this seminal work, Smith postulates the central thesis that a new paradigm of war, which he terms 'war amongst the people', has become the dominant form of conflict since the end of the Cold War. These conflicts differ fundamentally from conventional forms of interstate warfare and, as a consequence, contemporary armed forces of industrialised states face significant problems when engaging in this form of conflict. The key defining characteristics of 'war amongst the people' are their longevity and that they are fought between parties that are part of the population or which operate amongst the population. This puts such conflicts in stark contrast to the more prevalent conventional understandings of war, which portray war as a conflict between two, or more, uniformed armies that meet on a battlefield situated outside of population centres. In 'war amongst the people', the battlefield is no longer a clearly demarcated area that is dissociated from the population; instead, the population itself becomes part of the battlefield terrain.

While more widely applied, the term 'war amongst the people' has become extremely relevant in the British context following the unfinished campaigns in Afghanistan and Iraq. These campaigns had far-reaching impacts on the UK in general and the British Army in particular, in terms of human and material resources, international legitimacy and domestic politics.[12] For this reason, proposing a critical re-examination of 'war

---

(Stanford: Stanford University Press, 2008); Frank Kitson, *Low Intensity Operations. Subversion Insurgency and Peacekeeping* (London: Faber & Faber, 2011); Montgomery McFate, 'The Military Utility of Understanding Adversary Culture', *Joint Force Quarterly* 38:3 (2005); John Nagl, *Learning to Eat Soup with a Knife. Counterinsurgency Lessons from Malaya and Vietnam* (Chicago: University of Chicago Press, 2005); David Kilcullen, *Counterinsurgency* (Oxford: Oxford University Press, 2010); US Army, *Field Manual 3-24: Counterinsurgency* (Washington DC: Department of the Army, 2006).

10　John Nagl, 'Foreword' to Galula, *Counterinsurgency Warfare*, ix.

11　Frank G. Hoffman, *Conflict in the 21st Century: The Rise of Hybrid Wars* (Arlington: Potomac Institute for Policy Studies, 2007), 7.

12　James K. Wither, 'British Bulldog or Bush's poodle? Anglo-American relations and the Iraq war', *Parameters* 33:4 (2003), 67–82; Wither, 'Basra's not Belfast: the British Army, "Small

amongst the people', both in its conceptual and practical implications, seems almost an obligation for this first volume of the Sandhurst Trends in International Conflict series. If 'trends' are understood in the context of an attempt to apprehend directions of change, there is a need to understand how they have developed. As Colin Gray argues, 'The only sources of empirical evidence accessible are the past and the present; one cannot obtain understanding about the future from the future'.[13]

The remainder of this introduction is structured as follows: first, Smith's 'war amongst the people' paradigm is outlined; second, key debates surrounding this paradigm are considered; third, it is explained how this volume critiques and repositions Smith's paradigm as a prism for interrogating trends in contemporary conflict, with both conceptual and practice-oriented implications; and lastly, a synopsis of the contributions to this volume is provided.

The 'War Amongst the People' Paradigm

In the opening section of his work, Smith postulates that 'war no longer exists', at least not in the industrial form that was prevalent from Napoleon to the wars of the first half of the twentieth century.[14] These regular wars, which Smith terms 'industrial wars', militarised whole societies, made use of the entire resources of the modern state (wealth, population, means of production) and were aided by technological advancements that the industrial revolution made possible.[15] They saw regular armed forces in the service of the state fight an opposing armed force on a delineated battlefield until a decisive strategic victory was achieved. Of particular importance for industrial war is the decisive use of military force at the strategic level to achieve clear political objectives, which are generally accomplished through the destruction of the enemy's armed forces. As Smith noted, 'in what I call 'industrial war', you sought to win a trial of strength and thereby break the will of your opponent, to finally dictate the result, the political outcome you wished to achieve'.[16] A Clausewitzian understanding

Wars" and Iraq', *Small Wars and Insurgencies* 20:3/4 (2009), 611–35; Ben Clemens, 'Public Opinion and Military Intervention: Afghanistan: Iraq and Libya', *The Political Quarterly* 84:1 (2013), 119–31.

13   Douglas C. Lovelace Jr., Foreword to Colin S. Gray, *Thucydides was Right: Defining the Future Threat* (Carlisle Barracks: Strategic Studies Institute/US Army War College Press, 2015), v.

14   Smith, *Utility of Force*, 1.

15   Smith, *Utility of Force*, 31–107.

16   Toni Pfanner, 'Methods of Warfare: Interview with Sir General Rupert Smith', *International Review of the Red Cross* 88: 864 (2006), 719.

of war thus characterises industrial wars, where 'war has a dual nature: on the physical level it is a *trial of strength*, while on the psychological level it is *a clash of wills*. The trial of strength has primacy: destroy the enemy's war-fighting potential and thereby break its will'.[17] According to Smith, the two World Wars are the culmination point for this paradigm, which was then replaced by 'war amongst the people', which has become the defining feature of conflict in the present and likely future. Although the basic features of 'war amongst the people' can be found in historical conflicts, such as the Spanish guerrilla resistance during the Napoleonic Wars, he argues that it has only became dominant in the post-Cold War era. These wars differ fundamentally from conventional forms of warfare, specifically because, in 'war amongst the people', the Clausewitzian trinity – the unity of people, armed forces and government – effectively ceases to exist.[18]

In his outline of the 'war amongst the people' paradigm, Smith identifies six basic trends that characterise such conflicts. All six trends are fundamentally interrelated and are identified as being in constant flux, as their relative significance varies depending on circumstances unique to each conflict.

First, the ends for which wars are fought are changing. No longer are the hard objectives that decide a political outcome a decisive factor, 'but those that establish conditions in which the outcome may be decided'.[19] Whereas the conduct of industrial war was informed by clear strategic goals, post-industrial conflicts are characterised by attempts to 'create a conceptual space for diplomacy, economic incentives, political pressures and other measures to create a desired political outcome of stability, and if possible democracy'.[20] According to Smith, as the objectives of war are changing, so are the ways in which war is carried out. The crucial feature he identifies is the 'will of the people', which, in the traditional application of force, has often only played a subsidiary role, given the conventional focus on the destruction of the enemy (i.e. winning the 'trial of strength'). As a consequence, military victory in 'wars amongst the people' rarely leads to a political resolution, as Smith seeks to demonstrate using several case studies, including Algeria, Vietnam and Bosnia. This poses significant problems for contemporary armed forces of industrialised states, as 'it has become more difficult to translate the use of the military instrument into grand strategy and political success'.[21]

---

17 Kiss, *Winning Wars amongst the People*, 22.
18 Beatrice Heuser, 'Introduction' in Carl von Clausewitz, *On War* (Oxford: Oxford University Press, 2007), xxviii.
19 Smith, *Utility of Force*, 269–70.
20 Smith, *Utility of Force*, 272.
21 Rupert Smith, 'Thinking about the Utility of Force in Wars Amongst the People' in John Andreas Olsen (ed.), *On New Wars* (Oslo: On Defence and Security, 2007), 28.

Second, Smith emphasises that the theatre of war is the social space, where the people have become the battlefield. Conflicts thus take place amongst the people: 'in the streets and houses and fields – all the people, anywhere – are the battlefield'. As such, '[m]ilitary engagements can take place anywhere: in the presence of civilians, against civilians, in defence of civilians. Civilians are the targets, objectives to be won, as much as an opposing force'.[22] Furthermore, 'war amongst the people' represents a 'global theatre of war',[23] where military forces 'fight in every living room in the world as well as on the streets and fields in a zone of conflict'. As the people need to be won over, 'commanders and leaders alike need the media in order to … explain their own version of events. To this extent, the media is a crucially useful element in modern conflict for attaining the political objective of winning the will of the people'.[24] As capturing the will of the people is a political task that requires diplomatic, economic and military measures, armed forces alone are no longer able to deliver strategic objectives. Although still important, the utility of force in modern conflicts is diminished, as the military task (i.e. the capability to apply physical force) is only one supporting element in the conduct of 'war amongst the people'. Military force is therefore no longer strategic, but sub-strategic, as it 'is no longer used to decide the political dispute but to create the conditions in which strategic result is achieved'.[25]

Third, conflicts tend to be timeless, and, with victory more difficult to achieve, seemingly unending. Conflicts in the 'war amongst the people' paradigm are based on 'a continuous crisscrossing between confrontation and conflict, regardless of whether a state is facing another state or a non-state actor. Rather than war and peace, there is no predefined sequence, nor is peace necessarily either the starting point or the end point: conflicts are resolved, but not necessarily confrontations'.[26] Hence, '[t]he old cycle of cycle of peace-crisis-war-resolution-peace is being replaced by a different cycle: confrontation-conflict-confrontation-conflict, in which military force is one supporting instrument of reaching political goals by other means'.[27] 'Wars amongst the people' therefore do not lend themselves to 'quick fixes or solutions'.[28]

Fourth, Western forces have given ever-increasing primacy to force preservation. As Smith notes, 'We fight so as to preserve the force rather

---

22 Smith, *Utility of Force*, 3.
23 Smith, *Utility of Force*, 289.
24 Smith, *Utility of Force*, 286.
25 Pfanner, 'Methods of Warfare', 719.
26 Smith, *Utility of Force*, 181.
27 Kiss, *Winning Wars amongst the People*, 3.
28 Smith, *Utility of Force*, 189.

than risking all to gain the objective'.[29] This tendency has arguably resulted from the greater sense of casualty abhorrence that has characterised contemporary democratic societies (the so-called 'body bag' effect), the reduction in size and resources of Western armed forces as part of a post-Cold War 'peace dividend', as well as the financial cost of high-tech military armaments.

Fifth, '[o]n each occasion new uses are found for old weapons and organizations which are the products of industrial war'.[30] As these were constructed specifically for succeeding in interstate industrial war, they are of limited use in 'war amongst the people'. This has led Smith to controversially advocate that armies and organisations need to be adjusted to reflect this paradigm change, as 'the enemies we face today are of a completely different nature'.[31] In positing that enemies are no longer industrial states, but actors that rely far more on light weapons and suicide attacks, he argues that there is a need to adapt capabilities to these trends and not focus defence procurement on the necessities dictated by the age of industrial war.

Sixth, and last, in 'war amongst the people' non-state actors assume a central role.[32] Following Smith, interstate confrontations have (generally) given way to multinational intervention forces, in whatever form, be it coalitions of the willing or more formal alliances, being in conflict to a far greater extent with a range of non-state entities, such as insurgents, guerrillas and, in recent times, international terrorist networks.

'War Amongst the People': A Prism for Analysis?

While '[t]he "new wars" school of thought has contributed significantly to understanding why conventional military superiority has limited value in civil wars or counterinsurgencies',[33] it has not gone uncontested. Though it is beyond the scope of this introduction to conduct an exhaustive review, it is important to outline key criticisms. On the one hand, critiques have focused on whether these changes really constitute a permanent paradigm shift, with Rob Johnson noting that 'the scholarship of strategic studies provides several examples of paradigm shift that did not entirely live up to expectations'.[34] In a similar fashion, Colin Gray has highlighted the

---

29 Smith, *Utility of Force*, 271.
30 Smith, *Utility of Force*, 297.
31 Smith, *Utility of Force*, 299.
32 In this Smith's paradigm largely aligns with Mary Kaldor's 'New Wars' thesis.
33 Bart Schuurman, 'Clausewitz and the "New Wars" Scholars', *Parameters* 40:1 (2010), 89.
34 Rob Johnson, 'The Changing Character of War', *The RUSI Journal* 162:1 (2017), 6.

need to be 'acutely alert to the potential peril of confusing a few concrete cases with irrefutable evidence of a fully matured truth' when dismissing the utility of conventional military force in the twenty-first century, as 'we simply do not know what this century will bring'.[35] On the other hand, questions have been raised about whether these perceived changes are really something new or rather result from overlooking historical precedents and adopting an overly Eurocentric perspective.[36] Edward Newman has criticised the idea of 'novelty', arguing for a deeper historical analysis of past wars to discover how many of the elements identified by the likes of Smith in his 'new' approach really are so.[37] Indeed, as Antulio Echevarria claims, 'throughout history, terrorists, guerrillas, and similar actors generally aimed at eroding an opponent's will to fight rather than destroying his means'.[38]

From a practitioner's perspective, criticism has been raised in relation to the strategic, operational and tactical implications of re-shaping armies primarily in the direction of population-centric warfare.[39] For instance, Dan Cox and Thomas Bruscino lament that there is an 'implied and explicit violence aversion of the current population-centric approach',[40] while Gian Gentile claims that population-centric approaches 'may be a reasonable operational method to use in certain circumstances, but it is not a strategy'.[41] Steven Metz goes so far as to call for 'abandoning counterinsurgency'.[42]

Generally, these criticisms have emphasised continuity between old and so-called 'new' wars. More importantly, they have claimed that these changes do not equate to 'signifying a fundamental change in the nature of war'.[43] Given such a critique, it is understandable that practitioners

---

35   Colin S. Gray, *Hard and Soft Power: The Utility of Military Force as an Instrument of Policy in the 21st Century* (Carlisle Barracks: Strategic Studies Institute/US Army War College Press, 2011), 7.

36   Alexander McKenzie, '"New Wars" Fought "Amongst the People": "Transformed" by Old Realities?', *Defence Studies* 11:4 (2011), 569–93; Tarak Barkawi, 'Decolonizing war', *European Journal of International Security* 1:2 (2016), 199–214; Michael Lower, 'New Wars, Old Wars, and Medieval Wars: European Mercenaries as State Actors in Europe and North Africa, ca. 1100–1500', *Mediterranean Studies* 25:1 (2017), 33–52.

37   Edward Newman, 'The "New Wars" Debate: A Historical Perspective Is Needed', *Security Dialogue* 35:2 (2004), 179–85.

38   Antulio J. Echevarria, *Fourth Generation Warfare and Other Myths* (Carlisle Barracks: Strategic Studies Institute/US Army War College Press, 2005), 10.

39   Dan G. Cox and Thomas Bruscino (eds), *Population-Centric Counterinsurgency: A False Idol?* (Fort Leavenworth: Combat Studies Institute Press, 2011).

40   Cox and Bruscino, *Population-Centric Counterinsurgency*, 3.

41   Gian P. Gentile, 'A Strategy of Tactics: Population-Centric COIN and the Army', *Parameters* 39:3 (2009), 6.

42   Steven Metz, 'Abandoning Counterinsurgency: Toward a More Efficient Antiterrorism Strategy', *Journal of Strategic Security* 10:4 (2017), 64–77.

43   Schuurman, 'Clausewitz and the "New Wars" Scholars', 97.

have become concerned that any attempt to re-shape the strategic outlook of armies to fight these types of wars will lead to the diminution of conventional war-fighting capabilities. As Gentile has argued in relation to the US Army, 'fighting as a core competency has been eclipsed'[44] in the population-centric paradigm. This is particularly relevant during a time of renewed tensions between the US, China and Russia, which serve as a strong reminder of the potential consequences of the so-called 'tragedy' of great power politics.[45]

While these criticisms open important points for reflection, they do not damage the overall validity of Smith's promotion of 'war amongst the people' as a key feature of contemporary conflict. In the same way as Kaldor has referred to 'new wars' 'not as an empirical category but rather as a way of elucidating the logic of contemporary war that can offer both a research strategy and a guide to policy',[46] Smith's claims generate a critical (re)engagement with fundamental questions about war and warfare. This re-examination is particularly pressing as 'winning' has become a sort of chimera in contemporary conflicts. Rather than providing conclusive solutions, the 'war amongst the people' paradigm functions as a lens that provokes a series of questions concerning, amongst other things, the very nature of war, the identity of the actors involved and the long-term strategic impact of such conflicts on both the domestic societies of external intervening forces, as well as those in which the conflict takes place. These questions are not limited and bounded by one specific and defined type of conflict, thereby reflecting the 'increased merging or blurring of conflict and war forms'.[47] After all, cases such as Syria and Eastern Ukraine, are examples that escape easy categorisation.

In line with this, this volume treats the 'war amongst the people' concept less as a clear and fixed phenomenon and more as a conceptual and analytical prism through which contemporary conflicts can be questioned and critiqued. In comparison to other related concepts, such as intra-state conflict, asymmetric war and even counterinsurgency, 'war amongst the people' has the advantage of focusing ever greater attention on the essential element of 'the people'. Indeed, while war as a concept and as a phenomenon is a fundamentally human activity, the qualifier 'amongst the people' further specifies that, more than the alternatives, this type of warfare depends on and is defined by an extremely intricate network

---

44 Gian P. Gentile, 'Let's Build an Army to Win All Wars', *Joint Forces Quarterly* 52:1 (2009), 27.
45 John J. Mearsheimer, *The Tragedy of Great Power Politics* (New York: Norton, 2001).
46 Kaldor, 'In Defence of New Wars', 1.
47 Frank G. Hoffman, *Conflict in the 21st Century: The Rise of Hybrid Wars* (Arlington: Potomac Institute for Policy Studies, 2007), 7.

of different actors and dynamics. As a prism, it therefore can be used to illuminate these differing components, while simultaneously assessing them from multiple angles. Four such angles structure the contributions to this volume: the conceptual, the practical, the legal and the domestic. Although each are analytically divided into different components, they complement each other in re-composing the prism as a multi-focal lens for analysis.

This volume also works at the intersection of military practitioners and academic approaches, seeking to bring these into a wider and more productive conversation. As noted earlier, the development of the debate and literature on 'war amongst the people' has created some productive exchanges between academics and practitioner and this volume aims at furthering this cross-fertilisation. On the one hand, understanding 'war amongst the people' is central to the practice of operating in such environments, while on the other, practical operations are a central component in the analytical re-evaluation of the concept.

Synopsis of the Volume

The last section of this introduction presents a synopsis of each chapter in this volume, organised along the four angles of the prism. The first angle, the conceptual, is primarily an engagement with the concept of 'war amongst the people' itself, placing it under deeper scrutiny. The second angle focuses on the practical challenges originating from the paradigm. Pressing legal concerns emanating from such conflicts concern the third angle. The fourth and concluding angle of this volume focuses on the UK, evaluating the impact of 'war amongst the people' on the domestic context of the intervening state, both at the societal level and for the armed forces.

In the opening contribution to the volume, Beatrice Heuser invites us to take a 'step back' from 'war amongst the people' to 'reconsider what we think of 'war' itself'. Through the exploration of five interrelated questions about war, Heuser provides a historical overview of the concept of war, showing how its cultural, normative and strategic meanings have changed from early modernity to the present. At the heart of her analysis are the limitations of the human obsession with binary categorisations, as 'dualistic approaches, whatever their advantages ... make us blind to nuance and variation'. Phenomena such as war should therefore be placed on a sliding scale in order to capture their complexity. Drawing on historical examples, and in line with some of the earlier criticisms of Smith's approach, Heuser shows how war has always contained multiple elements:

Napoleon's Peninsular War contained classic army-against-army elements, and all forms of insurgency and counterinsurgency, while the Second World War contained important insurgency/asymmetric war elements, such as the Résistance or Soviet partisan warfare, alongside the air war, large-scale naval operations and the largest tank battles the world had seen.

As such, '[e]ven a single war can fit into different categories'. For this reason, she argues that reductionist, dualist categorisations of war should be abandoned in favour of a more complex level and scope of analysis along several spectra: 'at this stage, where on a spectrum is it in terms of the number of fatalities? Where is it in degree of popular support for the adversarial leadership? Where in terms of armaments, supplies, infrastructure and other resources? Where in terms of foreign support? Where in terms of war aims?'. Such a view allows a greater appreciation of war's peculiarities and its distinctive circumstances. Furthermore, as the understanding of war, like other phenomena, is strongly culture-dependent, any notion that peoples and cultures are interchangeable needs to be abandoned. In order to understand specific wars, their analysis should be aided by the expertise of country or regional specialists. From this, Heuser concludes that military officer training will 'have to include learning to work with country experts and with experts that can help in the analysis of that conflict. Officers need to learn to think of the right question to ask of them and to make sense of complex answers. That, in future, will be a key skill for any military commander'.

In the second chapter, Raitasalo takes up Heuser's point on war being culturally and contextually dependent, arguing that the 'war amongst the people' paradigm emerged out of Western, post-Cold War security thinking, which had identified "new' kinds of wars with Western militaries out-of-area' as a key priority. The development of this priority was caused by the demise of the Soviet Union and the threat-vacuum that was created. As such, 'Western statesmen and analysts of strategic affairs became troubled by the loss of solid foundations for planning and executing security and defence policy, including formulating guidelines for long-term military capability development'. It is within this context, in which Western states sought to redefine international security and reconceptualise war, that the 'war amongst the people' paradigm developed, as an attempt 'to cope with this post-Cold War era *problematique*'. Raitasalo argues that, while 'war amongst the people' identified tactical and operational level lessons that could be drawn from post-Cold War conflicts, it lacks sufficient engagement with the strategic level. Engaging with the strategic level would require analysing 'why Western states should manage intra-state wars around the world with their militaries'. Hence, while 'Smith's notion of the lowered utility of force in many contemporary conflicts is spot on', clinging to the 'war amongst the people' concept threatens 'Western security by engaging Western states in political and social crises that have little or no connection to their strategic security interests'. This lack of strategic awareness has implications beyond the mere engagement in these conflicts, as accepting the 'war amongst the people' thesis as a guideline for the development of military forces would lead to further atrophying of large-scale deterrence and war-fighting capability within the West in general and Europe particularly'. As these capabilities will be needed in the decades to come – the last decade has shown that great power politics is back with a vengeance – some of the prescriptions of Smith's paradigm need to be reconsidered and contextualised for the contemporary era.

Alex Waterman's contribution focuses on the notion of 'the people' within the overall paradigm. Waterman, in line with preceding contributions, highlights the complexity of conflict and war and similarly calls attention to the absence of constants, as each case shows difference over space and time. He illustrates this through the theme of 'the people', which not only enjoy centrality in Smith's paradigm, but in thinking on revolutionary and counter-revolutionary warfare more generally. As such, and in contrast to Raitasalo, Waterman identifies the intellectual basis for the 'war amongst the people' not in post-Cold War thought, but as 'rooted in the theories of insurgency and counterinsurgency (COIN) that emerged from late colonial and early Cold War rebellions'. Through a review of classical and contemporary COIN theory, Waterman argues that theorising about the nature of 'the people' in these works is situated within the Maoist tradition of insurgencies, which consolidated 'the role of "the people" as a core political element of insurgency theory'. COIN theory in general, and 'war amongst the people' in particular, creates an over-simplistic understanding of the people, which does not pay sufficient attention to 'the complex relations, cleavages, conflicts, coalitions and actors that make up the "people"'. Instead of a 'war amongst the people', Waterman introduces the notion of 'war amongst peoples', which encapsulates a broader range of counterinsurgent–population interactions than notions of 'the people' in Smith's contribution. Through an in-depth analysis of the Syrian Civil War and its plethora of actors, his chapter demonstrates the multifaceted nature of state–population interactions and highlights the often-overlooked socio-political nuances behind a conflict.

The next three chapters depart from a primarily conceptual analysis, focusing instead on the practical challenges these wars pose. Specific consideration is hereby given to the need for stronger attention to the actors fighting in 'wars amongst the people'. In Chapter 4, Whitney Grespin examines US efforts in 'supporting the development of the professional security forces in partner countries', which has become increasingly important in the contexts of Afghanistan and Iraq and as part of the Global War on Terror (GWOT) more broadly. Grespin critically evaluates the claim that this is a smart way to limit the expenditure of life and resources, as well as a way to limit the external military footprint of the intervening state. Contrary to this claim, she argues that 'the overall return on investment for these expenditures has been poor' and '[it] is difficult to quantify which partner capabilities have truly alleviated the operational requirements of US troops'. The resources that are allocated to the Security Force Assistance (SFA) programme are below the levels required to generate the desired institutional change. Grespin sees the need for a strong awareness of the deep political nature of any SFA programme, which 'reconfigures the authority structure in partner nations' and warns against the temptation of seeing the tactical military training of partner countries as a short-cut alternative to a more holistic strategic and operational approach, which include 'whole-of government approaches to nation-building'.

In Chapter 5, Vladimir Rauta examines the role of irregular forces in 'war amongst the people'. He specifically focuses on 'the strategic differences' between the role of proxies and auxiliaries. Rauta makes the case 'for differencing proxies from auxiliaries based on the former's politico-strategic role compared to the latter's military-tactical utility'. While proxies modify the original frame of political violence, auxiliaries conserve it, by simply offering tactical support to one of the warring parties in the conflict. While a proxy 'accounts for a shift in the structure of the conflict parties', an auxiliary 'does not alter the structuring of the conflict'. By examining the case of Afghanistan, Rauta shows how this conceptualisation allows for a more nuanced and multi-layered understanding of the conflict – one that captures the 'strategic interaction' of the parties involved, 'showing the variation of the many strategies of using armed third parties in civil wars'. This furthers understanding of the complex environment within which 'wars amongst the people' take place, which, following Rauta, is crucial, given that greater strategic clarity is essential for strategic success.

In Chapter 6, Georgina Holmes examines the pre-deployment training of female peacekeepers in Rwanda. Drawing on Pierre Bourdieu's theory of social practice, the chapter first shows that female peacekeepers are trained to be socialised in a specific gendered fashion and then considers the impact that this has on the practices of peacekeeping. Holmes illustrates how the pre-training environment is less focused on knowledge transfer, from trainer to trainee, but is a social space where gender norms are reproduced and contested. Drawing on interviews with military personnel, trainers and external consultants, Holmes makes the case that the training is too narrow in operationalising the United Nation's (UN) gender mainstreaming norms, which has led to a more limited preparedness of female peacekeepers when 'required to perform 'off script' as they so often do in real life scenarios'. This makes the enhancement of pre-deployment training essential 'for peacekeepers to learn the soft skills required for community engagement' in the context of population-centric war.

'War amongst the people' has raised a series of pressing legal questions. Originally designed for conventional interstate war, the Law of Armed Conflict (LOAC) has been forced to adapt to the ever-growing intra-state nature of contemporary conflicts. In her contribution on the legal aspects, Andree-Anne Melancon asks how 'war amongst the people' has affected the principle of distinction in LOAC. Specifically, she argues that the increasing blurring of the distinction between combatants and non-combatants requires a re-evaluation of the meaning of 'direct participation in hostilities (DPH)'. By drawing upon the debate between traditional and revisionist Just War traditions, Melancon argues for a less stringent interpretation of DPH to balance between the guiding moral principle of discrimination and its practical applicability in 'wars amongst the people'. By introducing elements of individual moral responsibility, Melancon argues that civilians can be considered 'legitimate targets' and 'are directly participating in the hostilities even if they are not taking part in the fighting directly'. Additional situational responsibility must also be considered in order to assess DPH. Melancon argues for a broader interpretation of DPH, which would include, 'inter alia, the fabrication of explosives, the funnelling of funds on the black market, transporting weapons, recruitment, and propaganda'. This, however, requires a more restricted 'temporal scope of application since it is more context-dependent'. The principle of discrimination remains thus applicable in the real-world context of asymmetric warfare.

In Chapter 8, Grant Davies integrates the conceptual discussion of Melancon's chapter with a practitioner's perspective on another central aspect of LOAC applicability in the context of 'war amongst the people': the issue of detention. As the recent campaigns in Afghanistan and Iraq have prompted strong scrutiny of detention procedures, he discusses the relationship between host state and assisting state in detaining policy and practice. Central to this discussion is how the right and responsibility to detain can be extended from the host state to the assisting state. The United Nations Security Council (UNSC) assumes a key role in this respect, but its influence is limited because contemporary military interventions have often taken place outside the UN's legal framework. Another key issue is the extent to which international human rights law should be applied in contexts of non-international armed conflict (NIAC). Davies argues that, at times, while international human rights law requirements can be onerous, 'there are huge political and presentational difficulties with any state derogating from Human Rights provisions'. A closer integration of international human rights law and international humanitarian law is necessary, according to Davies, to guarantee the rights of detainees, while also respecting the strategic and operational necessities of the host and assisting states. While the issue of detention in NIAC still has many grey areas, Davies concludes that 'a workable detention regime that fuses International Human Rights Law, International Humanitarian Law and the ability of states to conduct military operations effectively must be found'.

The last two contributions to this volume focus on an often-overlooked yet extremely important aspect of 'war amongst the people': the domestic context of the intervening state. In Chapter 9, Ian Wilson focuses attention on the centrality of the people 'at home' when engaged in 'war amongst the people' elsewhere. Concentrating on levels of British public support for recent conflicts, he examines how diverse theories may explain variations in this, concluding that, while 'there is no statistical correlation between casualty levels and public support', 'duration may be a more significant factor' in explaining the noted trends in public approval or disapproval. In effect, support declines as the war endures, with political leadership having little power to maintain consent for the conflict. By examining the case of Libya, Wilson suggests that 'success does seem to influence how the British public responds', as well as wider perceptions of legitimacy and moral considerations. For Wilson '[t]he public … seem to make a rational assessment based on the likelihood of success balanced against the legitimacy of the aim'. His findings have profound implications for policy makers wanting to maintain the support of their own people for 'wars amongst the people' overseas.

In the final contribution to this volume, John Bailey evaluates the doctrinal changes of the British Army that 'war amongst the people' has prompted. The aim of this chapter is to work as a 'useful illustration of factors affecting the production and use of doctrine for 'war amongst the people''. Bailey demonstrates that, while academic debate thrives on multiple and contrasting paradigms, writing doctrine entails bridging those with 'government policy and the pressing operational needs of practitioners'. The consequence is that 'doctrine is often a manifestation of compromise, rarely satisfying all parties'. The 'war amongst the people' paradigm has prompted the British military to overcome a doctrinal focus on a traditional state-centric approach to security and to include non-traditional security issues, particularly recognising the importance of human security, both at tactical and operational levels. This has prompted doctrinal changes, such as those outlined in the *Army Doctrine Publication: Land Operations,* which aims at crafting a dual role for the soldier: fighting the enemy as well as protecting the population. This last chapter also neatly highlights a shared concern with Heuser's opening contribution: the centrality of the relationship between academia and practitioners. While Heuser identifies the need for educating officers in the cultural and regional specificities of the conflicts they are operating in, Bailey shows that 'the role of academics in testing doctrine is vital in improving alignment between ends, ways and means'.

This volume seeks to advance both these concerns. Indeed, a conversation between scholars and practitioners, as will be highlighted particularly in the conclusion, can generate unique insights into contemporary conflict and illuminate the complexities that underlie it. By re-composing the prism, four key themes emerge out of the conversation between the different contributions to this volume: the socially constructed nature of 'the people', the paradoxical interplay between knowledge and practice, the link between strategy and tactics and the problem of establishing local and international legitimacy. Taken together, they point towards a destabilisation of the Clausewitzian relationship between politics and war in 'war amongst the people'.

References

Barkawi, T., 'Decolonizing war', *European Journal of International Security* 1:2 (2016).

Baudrillard, J., *The Gulf War Did Not Take Place* (Bloomington: Indiana University Press, 1995).

Caldwell, W.B. and Leonard, S.M., 'Field Manual 3-07, Stability Operations: Upshifting the Engine of Change', *Military Review* 88:6 (2008).

Cassidy, R.M., *Counterinsurgency and the Global War on Terror* (Stanford: Stanford University Press, 2008).

Christopher S. Chivvis, *Understanding Russian 'Hybrid Warfare' and what we can do about it*. Testimony before the Committee on Armed Services on 22 March 2017 (Washington DC: United States House of Representatives, 2017) available at https://www.rand.org/pubs/testimonies/CT468.html accessed on 21 December 2018.

Clemens, B., 'Public Opinion and Military Intervention: Afghanistan: Iraq and Libya', *The Political Quarterly* 84:1 (2013).

Cox, D. and Bruscino, T., (eds), *Population-Centric Counterinsurgency: A False Idol?* (Fort Leavenworth: Combat Studies Institute Press, 2011).

Davis, N., 'An Information-Based Revolution in Military Affairs', *Strategic Review* 24:1 (1996).

Der Derian, J., *Virtuous War: Mapping the Military-Industrial-Media-Entertainment Network* (Boulder, Colorado: Westview Press, 2001).

Echevarria, A.J., *Fourth Generation Warfare and Other Myths* (Carlisle Barracks: Strategic Studies Institute/US Army War College Press, 2005).

Galula, D., *Counterinsurgency Warfare: Theory and Practice* (Westport: Praeger Security International, 2006).

Gentile, G.P., 'A Strategy of Tactics: Population-Centric COIN and the Army', *Parameters* 39:3 (2009).

Gentile, G.P., 'Let's Build an Army to Win All Wars', *Joint Forces Quarterly* 52:1 (2009).

Gray, C.S., *Thucydides was Right: Defining the Future Threat* (Carlisle Barracks: Strategic Studies Institute/US Army War College Press, 2015).

Gray, C.S., *Hard and Soft Power: The Utility of Military Force as an Instrument of Policy in the 21st Century* (Carlisle Barracks: Strategic Studies Institute/US Army War College Press, 2011).

Gregory, D., 'War and Peace', *Transactions of the Institute of British Geographers* 35:2 (2010).

Hammes, T.X., *The Sling and the Stone: On War in the 21st Century* (St. Paul: Zenith Press, 2006).

Hartmann, U., *The Evolution of the Hybrid Threat, and Resilience as a Countermeasure* (Rome: NATO Defense College, 2017).

Hoffman, F.G., *Conflict in the 21st Century: The Rise of Hybrid Wars* (Arlington: Potomac Institute for Policy Studies, 2007).

Ignatieff, M., *Virtual War: Kosovo and Beyond* (London: Vintage Books, 2001).

Johnson, R., 'The Changing Character of War', *The RUSI Journal* 162:1 (2017).

Kaldor, M., *New and Old Wars: Organized Violence in a Global Era* (Cambridge: Polity Press, 1999).

Kaldor, M., 'In Defence of New Wars', *Stability: International Journal of Security and Development* 2:1 (2013).

Kilcullen, D., *Counterinsurgency* (Oxford: Oxford University Press, 2010).

Kiss, P.A., *Winning Wars amongst the People: Case Studies of Asymmetric Conflicts* (Lincoln: University of Nebraska Press, 2014).

Kitson, F., *Low Intensity Operations. Subversion Insurgency and Peacekeeping* (London: Faber & Faber, 2011).

Lower, M., 'New Wars, Old Wars, and Medieval Wars: European Mercenaries as State Actors in Europe and North Africa, ca. 1100–1500', *Mediterranean Studies* 25:1 (2017).

Mälksoo, M., 'Countering hybrid warfare as ontological security management: the emerging practices of the EU and NATO', *European Security* 27:3 (2018).

McKenzie, A., '"New Wars" Fought "Amongst the People": "Transformed" by Old Realities?', *Defence Studies* 11:4 (2011).

McFate, M., 'The Military Utility of Understanding Adversary Culture', *Joint Force Quarterly* 38:3 (2005).

Mearsheimer, J., *The Tragedy of Great Power Politics* (New York: Norton, 2001).

Metz, S., 'Abandoning Counterinsurgency: Toward a More Efficient Antiterrorism Strategy', *Journal of Strategic Security* 10:4 (2017).

Münkler, H., *Die neuen Kriege* (Reinbek bei Hamburg: Rowohlt, 2002).

Nagl, J., *Learning to Eat Soup with a Knife. Counterinsurgency Lessons from Malaya and Vietnam* (Chicago: University of Chicago Press, 2005).

Newman, E., 'The "New Wars" Debate: A Historical Perspective Is Needed', *Security Dialogue* 35:2 (2004).

Olsen, J.A., *On New Wars* (Oslo: Oslo Files on Defence and Security, Institutt for Forsvarsstudier, 2007).

Pfanner, T., 'Methods of Warfare: Interview with Sir General Rupert Smith', *International Review of the Red Cross* 88: 864 (2006).

Pinker, S., *The Better Angels of our Nature: A History of Violence and Humanity* (London: Penguin, 2012).

Schuurman, B., 'Clausewitz and the "New Wars" Scholars', *Parameters* 40:1 (2010).

Smith, R., *The Utility of Force: The Art of War in the Modern World* (London: Allen Lane, 2005).

US Army, *Field Manual 3-24: Counterinsurgency* (Washington DC: Department of the Army, 2006).

Wither, J.K., 'British Bulldog or Bush's poodle? Anglo-American relations and the Iraq war', *Parameters* 33:4 (2003).

Wither, J.K., 'Basra's not Belfast: the British Army, 'Small Wars' and Iraq', *Small Wars and Insurgencies* 20:3 / 4 (2009).

PART ONE

CONCEPTUAL DEBATES AND CRITIQUES

1

RETHINKING WAR

Beatrice Heuser

Introduction

This volume is dedicated to the reconsideration of the concept of 'war amongst the people', but this chapter takes one step back and reconsiders 'war' itself. It tackles five questions about war, touching on several paradoxes on the way.

Humans love to categorise and to try to impose analytical order on the world. Moreover, there is a peculiar human – possibly European – predisposition to conceptualise the world in dualistic terms (while the East Asian Yin and Yang are not as mutually exclusive, it seems). The limits of this categorisation, namely the line drawn between what is black and what is white, tend to be, if not entirely random or arbitrary, then at least strongly culture-dependent. This binary obsession has affected the way most thinkers approached the questions to be tackled in this chapter, even though reality can rarely, if ever, be reduced to such simple patterns. Indeed, reality usually consists of clusters of phenomena, with each individual case distinctive, even if, in many ways, it resembles others. At best, such phenomena can be placed on a sliding scale, while dualistic approaches, whatever their advantages (for example, to get a legal grip on a situation) are blind to the nuances and variation.

What is War?

In their love of categories, European cultures have tried to delimitate war from other forms of violent conflict. The results are usually dependent on the times and circumstances in which these categories were defined and are highly unsatisfactory once applied more generally.

This can be seen from the beginning, with the origins of warfare. For primitive cultures existing to this day, warfare is born out of cattle rustling and raids on the grain stores of neighbouring tribes. The disgust felt by those on the receiving end of such raids gave birth to the notion that raiding is robbery, but that military action – warfare – intended to retrieve stolen property or abducted humans was legitimate – see, for example, the origins of the Trojan War. It was a first conceptual step to distinguish between illegitimate, criminal raids on the one hand and (usually larger) operations aiming to recover stolen goods and people for the restitution of a *status quo ante* on the other, with only the latter being considered 'proper' wars. Nevertheless, small wars that were difficult to distinguish from armed robbery continued to exist, in Europe and beyond – for instance, the bands of *klephtoi* of Greece under Ottoman rule, who were smugglers and bandits one day and Greek independence fighters another, or insurgents the world over, driven to reliance on criminal networks to get the arms or other supplies that their state-adversaries naturally seek to deny them.

A distinction has been made since the seventeenth century, if not earlier, between war on a larger scale and lesser forms of conflict. The latter came to be seen since Antiquity as private war as opposed to public war, or as a family feud on the one hand as opposed to a public war sanctioned by a higher authority on the other. The history of the European Middle Ages – and indeed of Early Modern times – revolves in good part around attempts to monopolise the use of force within states and thus to ban private war (although trends towards privatising elements of war in the more contemporary era are evident, as discussed later in this volume).

A third distinction has influenced academic thinking in this area: Plato distinguished between war among the Greek tribes and cities, which was called *stasis*, and war between Greeks and foreigners, primarily Persians, which alone he thought deserved the proper term for war: *polemos*. Adapting it to the construct of inter-state relations as the legitimate realm of violent conflict and the State as a zone of peace, this distinction has been applied to legitimate war versus insurgency, the idea of rebellion against authority, which implies an uprising against the established order of things. In European cultures, insurgencies were almost invariably criminalised, even though some cultures articulated a rationale for tyrannicide or for the fight for freedom (or national independence), thus seeking legitimation for their rejection of tyrannical regimes.[1] From the Romans to the Russians today, regimes in Europe have

---

1   A great example of this is the insurgency of the Thebans (in the name of liberty) against Macedonian rule, cruelly repressed by Alexander III, in 336 BCE.

sought to portray and treat rebels as criminals, whether they be slaves or Gauls or Chechens, rather than as legitimate combatants and have treated them quite differently – with far greater brutality – when caught than they treated adversaries defined as legitimate. The legal distinctions between legitimate forms of war and illegitimate forms depend heavily on who defines what is legitimate i.e. on *who has the power* to define what is legitimate and what is not.

All three sets of binary categorisations are heavily dependent on the worldview of their day and age i.e. on historical context. All three, seen retrospectively, are somewhat arbitrary (if politically motivated) impositions of distinctions on a reality that presented cases of all shapes and sizes, along sliding scales on which no clear lines are drawn by nature. Definitions of war made in *one* context rarely fully apply to another, as so many circumstances change. There seems to be a convergence of opinion that many, perhaps all, cultures have a notion of something large-scale and violent that creates disorder and destruction in the extreme and in which (at least) two sides are pitted against each other, namely war. However, the specific form of war that people had and have in mind when defining is context-specific and depends very much on what they know and often what they have seen in their own lifetime. This context-specific definitional process is worth bearing in mind when assessing the longevity and relevance of concepts such as 'war amongst the people'.

Primatologists or archaeologists will see evidence of 'war' when flanges of 40 apes give battle with another tribe of apes or when a mass grave of 30 skeletons is found that have met with a violent death. While at Thermopylae in 480 BC, when Leonidas and the 300 encountered the Persians – and assuming that they killed twice as many Persians as they numbered themselves – the total number of battlefield casualties would still not have qualified this epic historic battle as 'war', according to the reckoning of the Correlates of War project. That defines war as something with at least 1,000 battlefield deaths. Yet a battle with 1,000 deaths is closer, quantitatively, to the fatalities incurred in cattle rustling and gang warfare and is still closer to the death toll of the Battle of Thermopylae[2] than it is to the so-called *battle* of the Somme, which lasted for months and saw the death of over 400,000 men.[3] The curators of the Correlates of War project have, in the meantime, identified many of its shortcomings and

---

2   Admittedly I am cheating a little here, as the Second Persian War, of which the Thermopylae battle was the only major land battle, was decided by a naval battle with unknown numbers of casualties, and these would probably have lifted the war across the Correlates of War threshold.

3   For details, see the Correlates of War Project at http://www.correlatesofwar.org/ accessed on 10 October 2018.

adjusted their data set.[4] The problem remains that war comes in a myriad of forms (Clausewitz with his chameleon comes to mind). Arguably, there are clusters of wars that resemble each other more closely, but not merely two, and all attempts to categorise will do injustice to borderline cases. Few generalisations will easily apply to all of them. Clausewitz claimed that each age and civilisation has its particular form of war, but there is a need to go further and recognise that, within one age, within one civilisation, there can be several different forms.

On one end of this vast spectrum of war with its many permutations, there is war as conceived by Sun Tzu. It has something in common with some medieval warfare, perhaps even with some Early Modern wars. Yet, his advice to leave the enemy unsure as to where you want to give battle, or to withdraw quickly only to resurface elsewhere, makes *no* sense when the war is between two states with armies of tens or hundreds of thousands and the aim is to hold or reoccupy a well-defined piece of territory. Clausewitz, in turn, found it difficult to escape the paradigm of the Napoleonic Wars and to come to a more general view of war, although he tried to do so in the last years of his life. Some of what he wrote in *On War* is thus not useful or applicable to all wars, but applies to the extreme opposite end of the spectrum of wars from that covered by Sun Tzu.[5]

Paradox 1: Chaos vs Rules

A derivative of the European obsession with mutually exclusive binaries is the paradox. It stems from the realisation that such binary patterns are problematic. The essence of 'war' and 'guerre' is well captured by its etymological derivation from the Germanic word 'Wirren', meaning chaos, disorder or confusion. In binary terms, disorder and confusion are the opposite of order, rule-bound behaviour and the rule of law. It is at this point that the first paradox becomes clearer: much thinking and writing about war revolves around attempts to impose rules on what is the extreme of unruly behaviour. The rules invariably are restraints, in some form, or limitations on violence.

So, while war means confusion, disorder and chaos, many wars are fought according to rules and with constraints. Through explicit or implicit mutual agreement, perhaps through unilateral choice, perhaps following the laws of war, individual buildings, such as hospitals, or entire areas

---

4   See http://www.correlatesofwar.org/news/a-new-dyadic-war-dataset-is-published accessed on 10 June 2018.
5   For details, see Carl von Clausewitz, *On War* (Oxford: Oxford University Press, 2008) and Sun Tsu, *The Art of War: The Ancient Classic* (New York: Capstone, 2010).

may be defined as constituting sanctuaries that must not be bombed or otherwise destroyed. Indeed, the laws of war – *jus in bello* – impose many rules and restrictions on the chaos that is unleashed. Beyond specifically defined rules of war, there are also customs of war, the breach of which is usually seen as particularly shocking and escalatory. Such a breach of the rules might be the mutilation of corpses (instead of their honourable burial), the deliberate or accidental targeting of civilians, the destruction of crops or the injuring of soldiers in ways seen as particularly unacceptable – perhaps dishonourable – in that particular culture. Thus, patterns may emerge from war which, at first sight, seem curious to an outsider, such as that a warning is uttered before particular places are bombed or that both sides refrain from particular acts of war while threatening to resort to them as a form of escalation or punishment. Wars are thus – in theory if perhaps less so in the infinite variety of practice – subject to certain rules of engagement.

Paradox 2: Violence vs Discipline

Directly connected with this paradox of chaos and rules of war is the paradox of the armed forces as disciplined instruments of violence, not to be unleashed in an uncontrolled way, but in a focused manner against specific targets with specific purposes. This, at any rate, is what is supposed to characterise armed forces as an instrument of an orderly state based on the respect for law – from international humanitarian law to human rights law and indeed the law of government that will be imposed on an occupied area. The problems of restraining force and directing it are at the heart of the work of any military. Not all war is thus regulated – there are wars that see unbridled combatant violence deliberately unleashed.[6] Such violence is difficult to subordinate to a particular political purpose, let alone to terminate under conditions that can be turned into lasting peace.

The paradox for the armed forces is also that, especially during 'wars amongst the people', they may be called upon at different stages or in very short succession to fulfil utterly contradictory duties, from the infliction of extreme violence to extreme restraint. They may even be required to perform duties of imposing order on chaos – admittedly coupled with the threat of violence – in ways converging with those of police forces. To illustrate this, US General Charles Krulak coined the buzzwords 'strategic

---

6   For instance, the Duke of Alba's campaign in the Dutch War of Independence and the Spanish in the West Indies, the Germans on the Eastern Front and the Japanese and Soviet mass-rape in the Second World War.

corporal' and 'three-block war',[7] although this poses the question of when such activity is to be considered a police action and not warfare, as both can serve the enforcement of law. This brings us to the question of when reference can be made to the *state of war.*

When Is There a State of War?

In European cultures that codified warfare as something that had to obey rules, war and peace were seen as distinct, with the transition formalised and marked by elaborate ritual. The Roman republic developed such a ritual and, by the time Cicero wrote in the late first century BCE, it was one of the criteria of fighting a just (or proper) war that it had thus to be declared formally.[8] In Medieval Europe, heralds were dispatched to the enemy to serve notice of the declaration of war. With the printing press began the habit of declaring war in writing and explaining the reasons in print, distributing this, together with terms and conditions for neutrality of third parties, throughout Europe. The Ottoman sultan by contrast would declare war by displaying a number of *tuğlar* – tall poles with horsetails at their top, the Turkish equivalent of banners – in front of Topkapı Sarayı at Konstantinyya, his main residence.[9]

In practice, the form was not always respected. Already, in the eighteenth century, the habit of going to war without formal declaration was widespread. Between 1700–1870, only ten cases of wars between European powers were declared, while 107 cases of war began without a formal declaration.[10] The issuing of an ultimatum – already practised by the Romans – can replace the declaration of war, so that the triggering of the state of war can be blamed on the enemy, who refuse to satisfy the terms of the ultimatum by a certain deadline.

Declarations of war have the advantage that they leave no question mark as to the applicability of the *jus in bello,* suspending the normal laws of society that, of course, include the proscription of homicide. Homicide, in times of war, is permissible if it furthers the war aim and can be presented as proportionate to and necessary for the pursuit of the overall war aim. Thus,

---

7   C.C. Krulak, 'The Strategic Corporal: Leadership in the Three Block War', *Marines Magazine* (January 1999), available on http://www.au.af.mil/au/awc/awcgate/usmc/strategic_corporal.htm accessed on 8 June 2017.

8   Cicero, *De Officiis* (Boston: Loeb Classical Library at Harvard University Press, 1989), I.34–8.

9   Andrew Wheatcroft, *The Enemy at the Gate: Habsburgs, Ottomans and the Battle for Europe* (London: Pimlico, 2009), 13ff.

10 See J.F. Maurice, *Hostilities without Declaration of War from 1700 to 1870* (London: Her Majesty's Stationery Office, 1883).

by declaring 'war on terror', the George W. Bush Administration in the US could justify this and many other measures that would be unacceptable in peacetime. Yet it arguably conferred a quasi-legitimacy on terrorists as combatants and was heavily criticised for this reason.[11] The term was abandoned in the US under President Obama but lives on in France, justifying a state of emergency and the deployment of soldiers to guard France's domestic security after a series of terrorist attacks (a taboo in many countries). In the end, it is *primarily* a subjective, political decision, not an apolitical and objective evaluation of factors that determines whether one is or considers oneself to be 'at war'.

Just War or War as a Duel?

There were noted early attempts to create distinctions between an illegitimate 'bad' and a legitimate 'good' side in war: the cattle rustlers or the Trojan abductors of Helen of Sparta were cast as the 'bad' side, while those who set out to reclaim their property were the 'good' side, prosecuting a 'just' war. Plato attributed to his teacher Socrates acceptable reasons for going to war: one might usefully claim to be the 'victim of deceit or violence, or spoliation'.[12] Plato's own disciple Aristotle went further and listed 'arguments for making war on somebody'. These included that one or one's allies had been wronged.[13] By the time Cicero was writing, this had apparently become a well-established notion in Roman legal thinking and the conditions for a Roman declaration of war on an enemy was that it had to come 'after a formal demand for restoration', implying that this had to be in response to an injury received.[14] In theory at least, war should thus not be waged merely to further Rome's ambitions.

After the early Christian thinkers had torn themselves and each other apart over the question of whether 'thou shalt not kill' did or did not rule out war, majority consensus was established that collective self-defence – along the lines put down by Cicero – *was* legitimate. Popes and sovereign princes claimed the right to pronounce on whether this condition was obtained. Medieval and Early Modern authors on war tended to preface their works – or discuss at length – the need for war on one's own side to

---

11   See, for example, Mary Ellen O'Connell (ed.) *What is War? An Investigation in the Wake of 9/11* (Leiden: Martinus Nijhoff, 2012).

12   Plato: *Alcibiades* – quoted in Cian O'Driscoll, 'Rewriting the just war tradition: just war in classical Greek political thought and practice', *International Studies Quarterly* 59:1 (2015), 1–10.

13   Aristotle, quoted in ibid.

14   Cicero, *De Officiis*, I.36.

be legitimate, namely to fulfil the Graeco-Roman conditions for being a just war that one had been wronged.[15]

Yet there was a rival strand of thought that can also be traced to Cicero, who claimed that the Latin word *bellum* had its etymological origins in the word *duellum,* namely the duel.[16] That war may be the clash of two equal sides – even two *morally* equal sides – is a subject that has been debated since Antiquity. Breaking with the Christian just war tradition, in the fourteenth century, Giovanni da Legnano addressed the problem that, in a war, *both* sides might believe themselves to be justified in waging a particular war.[17] Theorists of sovereignty also tried to ignore the question of right or wrong. Discussion of just war was fully eclipsed for a century from Clausewitz onwards, when a new consensus was established that might was right and that the sovereign state could do as it pleased. This led to a new departure also in international law, in the law of the right to go to war, *jus ad bellum.* Lawyers of the nineteenth century tried to work their way around the new approach by arguing that it mattered not who went to war or why.[18] Instead, they concentrated their efforts on trying to limit the miseries inflicted by war. In the twentieth century, following the adoption of the Kellogg-Briand Pact of 1928, going to war can only be presented as legitimate if it takes the form of self-defence, while, to defend against aggression, one does not declare war but merely reacts to a state of war imposed by the aggressor. Consequently, the formal declaration of war has become even less regularly observed.

In popular perception, however, the dualism of the 'good versus evil' view in war continued to exist, especially in religious and ideological wars. After a slump from 1648 until 1792 these regained massively in strength on another level since the French Revolutionary and Napoleonic Wars, as wars turned nationalist, with nationalism being the new religion, unleashing hatred to degrees unseen since the religious wars. One might have thought that, after 1945, nationalism was overcome by a more enlightened view of the world, based on embracing shared values of human rights and democracy in the liberal West and by the denial of nationalism in the Communist view of class struggle in the Socialist world. After 1991, however, the world has borne witness to the resurgence of

---

15 See, for example, Anthony F. Lang Jr, Cian O'Driscoll and John Williams (eds), *Just War: Authority, Tradition, and Practice* (Washington DC: Georgetown University Press, 2013).

16 Cicero: *Brutus, Orator,* 153 (Boston: Loeb Classical Library at Harvard University Press, 1939), 426.

17 For details, see Giovanni da Legnano, *Tractatus de Bello: Represaliis et Duello* (London: Andesite Press, 2017).

18 Stephen Neff, *War and the Laws of Nations: A General History* (Cambridge: Cambridge University Press, 2005), 201.

nationalism in Eastern Europe, and more recently, in the US and Western Europe. States outside Europe, such as India or many Arab states, which did not go through the bloody catharsis of the Second World War, had never abandoned the nationalist dualism of 'us' and 'them' that has infected the whole world at some point, since it began to spread from its hearth in Europe in the early nineteenth century. Also, while Europe since 1945 seemed to have found consensus in secularism, Islam never settled for rendering to Caesar what is Caesar's and to God what is God's. Thus religion is not dead as an interpretive prism and fuels binary perceptions of friend and foe in the Middle East and beyond.

What are War Aims? What do People Fight For?

People tend to articulate their own feelings, aims and ambitions in terms and categories they have been brought up to find acceptable and appropriate. At times, there have been difficulties identifying what people aimed to achieve by going to war, given that some of their motivations may never have been articulated, perhaps not even to themselves.

Leaving aside straightforward raiding warfare, such as that conducted by Genghis Khan or the Huns, a variety of war aims has been identified over time. For polities that have lived and continue to live alongside each other war is often conceptualised as caused by a dispute and, in this context, justice is evoked, usually by both sides, as recognised by thinkers on the subject since Giovanni da Legnano. However, the narrow causes of wars – the spark or even the powder keg – are not the only determinants of war aims, which indeed may develop and mutate over the time of the war.

War aims range from very limited – the freeing of hostages or the recovery of stolen property – to very far-reaching: conquest of land, enslavement or even extermination of a population or the imposition of one's own values, laws or religion on the rest of humanity. Greek and Roman wars mainly revolved around the conquest and defence of crucial towns or around a frontal battle, on land or at sea. This is what they prepared their armies for. The convention demanded that the defeated party withdrew from the battlefield and accepted its defeat. Conditioned by such expectations, cultures drawing on Greek and Roman literature defined similar aims and, like the Ancients, found it very annoying if a defeated enemy did not admit defeat but continued to fight, in an 'irregular' way, even after they had lost the 'regular' *Hauptschlacht* – the supposedly decisive battle. Accordingly, such irregulars would be treated much more harshly. They would be punished, in fact, for refusing to submit to what

Pufendorf referred to as the 'arbitrament of Mars'[19] and for refusing to accept the outcome of a battle as the settlement of the dispute in favour of one side.

Turning to siege warfare, admittedly there is a certain timeless logic in the hate-driven aim of devastating an enemy's dwellings. Yet, whether there *was* one capital town that, if occupied, would lead to the fall of the leadership or the resignation of the government and the collapse of resistance depends very much on the context of the times. The aim in the 2003 Second Gulf War was to take Baghdad and bring down Saddam Hussein. Here again, even if the expectation of the ensuing collapse of the polity is widely shared, the vanquished party can refuse to accept defeat: its government can withdraw to another place from which, symbolically, to continue the resistance of the polity, as several state governments did when the Germans overran their country in the Second World War.

Of course, war aims are not driven *exclusively* by culture. Means and geography are also essential factors. These factors are interrelated. However, there is also an element of decision. How are war aims articulated and agreed upon? This seems to depend heavily on the imagination of those that do the defining and on the culture they share with those whose support they seek for the enterprise. Those who define and agree on war aims are usually an elite. It is astonishing to what extent such elites managed to garner the support of illiterate masses for loftier grand strategic aims they can barely have understood.

War aims are often defined in the light of tools available to wage war. One may try to kill a spider with a hammer, because one *has* a hammer, but not insecticide. Yet it is not a universally shared assumption that one has to swat the spider, rather than catch it with a net and throw it out of the window. With some conflicts, the very notion that they have to be resolved through war may be culturally dependent; in fact, some cultures seem to *need* bloody conflicts to perpetuate their own values. Indo-European cultures – but others as well, such as the early modern Japanese or every anthropologist's favourite, the Yanomamö – have developed around ideals of the warrior, rather than ideals of the merchant-mediator or the philosopher-pacifier. And warriors cannot be warriors unless there is war. Many repressive systems have needed an external enemy to excuse the repression of freedom – an idea most persuasively expressed in Orwell's *1984* and in Christa Wolff's *Cassandra*, both fables based on the realities of the Cold War. That conflict is a zero-sum game, incidentally, is also an ideologically-, and thus a culturally-dependent, assumption.

---

19   Quoted in Neff, *War and the Law of Nations*, 131.

Nevertheless, pretext or not, reasons put forward must be such as those who are to do the fighting can associate with and which seem worth their own blood and life. This, incidentally, is where the so-called 'Realist' school of International Relations theory is absurdly misleading: few people will be aroused merely by the promise of loot or the conquest of territory. Humans like to see themselves as moral and ethical, not as greedy and base. They like to turn even the basest of instincts into an uplifting and ethical motivation. In this context, it is necessary to briefly consider the 'greed versus grievance' debate, which, again, is a fallaciously binary interpretation applied to a multifaceted reality. In most cases, both motives can be found, plus multiple other factors, such as reputation and honour, the desire to avenge a dear friend or relative or to please a vengeful, monstrous deity by killing those who displease it. Externalising one's own frustrations by killing humans defined as horrible heathens, as *kafir* or, worse, heretics and renegades, or people who have been defined as sub-human and yet as the dangerous enemy to one's own society and prosperity, seems to give humans great satisfaction.

In wars that drag on, war aims can change considerably over time. This can be a function of various factors – for instance, mounting numbers of casualties can sap any readiness to settle for a compromise or, contradictorily, can lead to war weariness and the urge to cut one's losses. Other variables include aspects of enemy behaviour, enemy war aims, the demands of allies and the internal discourse about war aims (such as to create 'a world fit for heroes'). While the initial response of an attacked group may be just to defend itself, the war might unleash forces arguing not only for the restoration of the *status quo ante* but for lasting, structural change, from the constitution of a 'new world order' to the extermination of the threat at source (today usually expressed through some variant on demilitarisation and democratisation).

Constructs are prominent when it comes to war aims. Since Antiquity, the aim of 'liberty/liberties' or 'freedom' has been used as a catch-all which, when unpacked, contained amazingly different assortments of demands, while often lacking demands which could be viewed as essential to 'freedom'. To take other war aims, from today's perspective, it is difficult to follow how, in the past, people would fight for an infant's right to be king or over the *iota subscripta* in the Nicene Creed. From a twenty-first-century Western perspective, where the *status quo* is largely seen as satisfactory, it is unintelligible how, in other societies, people might fight to establish a religious dictatorship, as in Iran or the Islamic State (IS). War aims have nominally included constructs that were never even remotely realistic, such as to obtain 'command of the sea' or to establish or supposedly even restore a global 'universal monarchy'. And war aims beyond the planning of the central authority can be articulated, with or without serious consequences: in the First World War, French soldiers wrote 'à *Berlin*' on railway wagons as they left for the front and German soldiers sang '*wir fahren gegen Engelland*' (implying an invasion of England).

How Do Wars End?

A war can extend to the far corners of the globe or it can remain a geographically narrowly confined affair or even a mixture of both. It can escalate beyond anything that the parties involved had imagined rather than end in a nice, neat 'blitzkrieg' victory or it can be over in a few days. Length does not determine intensity – a four-year war such as the American Civil War may shed more blood than the more recent long, drawn-out Western interventions in Iraq and Afghanistan. Expectations about war aims and outcomes, but also the way in which a war has been fought, are clearly also *a* or *the* central factor in how much effort each side makes to win (or not to lose) and thus how wars end and how long they take. Think of the Japanese determination in the Second World War to fight on and on (or even to commit mass suicides, as on Okinawa), based on the quite mistaken expectation they had of the behaviour of US soldiers.

Expectations about how wars end are also linked to (at least the victor's) expectations about what *should* happen after it has ended. What situation *post-bellum* can be regarded as stable i.e. when can peace be considered to be 'lasting'? Death and destruction are the obvious consequences of wars. Other consequences can include the integration of the defeated country and people into the realm of the victor or a revolution or the long-term impoverishment of a region. Alternatively, the ascendency of a power, bringing modernisation in its wake and rising to be a new hegemon may be the consequence. Colin Gray has stressed that wars can decide big, smouldering conflicts and end them definitively,[20] while Ian Morris stresses the long-term benefits for defeated (by implication: culturally inferior) societies derived from being integrated into the realm of the victorious powers.[21] A weighing of bad and long-term beneficial effects of war for those who survive it has to include the ethical question of whether the greater happiness of the many should be bought by the terrible suffering of the few, relatively speaking. Today, even 'few' can mean tens of millions.

---

20 See, for instance, Colin S. Gray, *Fighting Talk* (Santa Barbara: Praeger Security International, 2007).

21 Ian Morris, *War: What Is It Good For? Conflict and the Progress of Civilization from Primates to Robots* (New York: Farrar, Strauss & Giroux, 2014).

Since Antiquity, arguably the most legitimate war aim is the restoration of the *status quo ante*,[22] but, in reality, this is impossible. Neither can the dead be brought back to life, lost limbs replaced, nor can all displaced populations be persuaded to return to their places or origin and their old lifestyles, nor can reconstruction restore the art and architecture that have been destroyed. No war will leave the culture in which it took place untouched. At the end of war, all sides will inevitably have lost much life and treasure and even the *guerre zéro morts* – the war in which one's own side has zero combat deaths – will bring much devastation to the other side. How one gets out of such wars, with the hatred such loss engenders where it does not already exist, is a challenge of the first order, every time. Pretending that this will work without social engineering or 're-education' is pointless. Even there, different historical wars have produced very different outcomes, ranging from the 'cabinet wars' of the Ancien Régime, few of which seem to have engendered lasting hatred among the population, to the immense hatred engendered by nationalist wars and civil wars with an ethnic and / or religious component. The latter would often poison the lives of later generations.

Conclusion

Everything discussed until now indicates that the answer to Clausewitz's injunction to establish 'what sort of war' one is engaging in is not an easy one, and this is without touching on the dynamics of war that can transform any conflict. Thus a terrorist act – the assassination of a prince in Sarajevo – or a limited war – the invasion of Poland – could turn into a world war. The most recent Iraq war was quite different in 2003 from what it was in 2006 or 2009. Wars that last more than a few days are usually analysed in terms of phases – a blitzkrieg phase, an insurgency phase, a phase where an outside power intervened in force – and thus defy any dualist categorisation. Others are many things simultaneously – to name but two examples, Napoleon's Peninsular War contained classic army-against-army elements and all forms of insurgency and counter-insurgency, while the Second World War contained important insurgency / asymmetric war elements, such as the *Résistance* or Soviet partisan warfare, alongside the air war, large-scale naval operations and the largest tank battles the world had seen. Even a single war can fit into different categories.

If there is anything to be taken from this confusion, it is to abandon dualist and excessively reductionist categorisations of war. Instead, an

---

22 See O'Driscoll, 'Rewriting the just war tradition'.

armed conflict one is facing should be analysed along several spectra or axes: at *this* stage, where on a spectrum is it in terms of the number of fatalities? Where is it in degree of popular support for the adversarial leadership? Where in terms of armaments, supplies, infrastructure and other resources? Where in terms of foreign support? Where in terms of war aims?

It means also homing in on its peculiarities, its distinctive circumstances, best explained by a country or regional specialist, and the abandonment of the crass idea that peoples and cultures are interchangeable. It was nonsense to think that one could 'work with the tribes' in Afghanistan as it had seemed to work in Iraq, as the two tribal systems are very different.[23] This also means that, in terms of educating officers, there remains the challenge of teaching them not only about 'major' or 'small' war but of preparing them for anything in between, perhaps all at once. A wider understanding of military education has come a long way from mindless drill or even of war by field manual, war by standard operational procedure or war by rote. In the more complex contexts of 'war amongst the people', which includes *peoples* with all their respective cultures and traditions, intervening forces are forced to engage with the populations in a myriad of ways. No officer can be expected to be the country expert for all countries that may become the theatre of operations. Officer training will thus have to include learning to work with country experts and with experts that can help in the analysis of that conflict. Officers need to learn to think of the right question to ask of them and to make sense of complex answers. Regardless of the type of conflict faced, that, in future, will be a key skill for any military commander.

References

Cicero, *Brutus, Orator* (Boston: Loeb Classical Library at Harvard University Press, 1939).

Cicero, *De Officiis* (Boston: Loeb Classical Library at Harvard University Press, 1989).

von Clausewitz, C., *On War* (Oxford: Oxford University Press, 2008).

Correlates of War Project at http://www.correlatesofwar.org/, accessed 10 October 2018.

Gray, C.S., *Fighting Talk* (Santa Barbara: Praeger Security International, 2007).

---

23 Montgomery McFate, *Military Anthropology: Soldiers, Scholars and Subjects at the Margins of Empire* (London: Hurst, 2017), 209.

Krulak, C.C., 'The Strategic Corporal: Leadership in the Three Block War', *Marines Magazine* (1999).

Lang, A.F., O'Driscoll, C., and Williams, J. (eds), *Just War: Authority, Tradition, and Practice* (Washington DC: Georgetown University Press, 2013).

da Legnano, G., *Tractatus de Bello: Represaliis et Duello* (London: Andesite Press, 2017).

McFate, M., *Military Anthropology: Soldiers, Scholars and Subjects at the Margins of Empire* (London: Hurst, 2017).

Maurice, J.F., *Hostilities without Declaration of War from 1700 to 1870* (London: Her Majesty's Stationery Office, 1883).

Morris, I., *War: What Is It Good For? Conflict and the Progress of Civilization from Primates to Robots* (New York: Farrar, Strauss & Giroux, 2014).

Neff, S., *War and the Law of Nations* (Cambridge: Cambridge University Press, 2005).

O'Connell, M.E. (ed.), *What is War? An Investigation in the Wake of 9/11* (Leiden: Martinus Nijhoff, 2012).

O'Driscoll, C., 'Rewriting the just war tradition: just war in classical Greek political thought and practice', *International Studies Quarterly* 59:1 (2015).

Sun Tsu, *The Art of War: The Ancient Classic* (New York: Capstone, 2010).

Wheatcroft, A., *The Enemy at the Gate: Habsburgs, Ottomans and the Battle for Europe* (London: Pimlico, 2009).

2

THE UTILITY – OR FUTILITY – OF FORCE?

What is Wrong with the 'War Amongst the People' Thesis?

Jyri Raitasalo

Introduction

The nature and character of war has attracted lot of scholarly attention within the disciplines of International Relations, Strategic Studies and Security Studies, to name but a few. A majority of scholars agree that, while the *nature* of war changes slowly, if at all, the *character* of war – the detailed manifestation of war at any *particular era* – is more subject to change.[1] Rupert Smith's *The Utility of Force – The Art of War in The Modern World* is one of many recent attempts to come to terms with the problem that Western states in general and the US in particular have faced during the last 25 years. This problem is related to the many military operations that the West has executed since the Cold War's demise and the perception that most Western military operations have not produced positive results. On the contrary, on several occasions these operations have made the situation worse, with the examples of Iraq (2003–) and Libya (2011) troubling cases in point.

The Utility of Force describes eloquently the long lineage of change within the character of war. From the end of the Second World War until the end of the Cold War, the number of civil wars or intra-state

---

1 On the nature and character of war, see, for example, Colin S. Gray, 'How Has War Changed Since the End of the Cold War?', *Parameters* XXXV:1 (2005), 14–26.

wars increased. A vast majority of armed conflicts – wars – today, and in the preceding decades, have been intra-state rather than interstate.[2] Furthermore, during the post-Cold War era, expectations for the West to do something to decrease or contain the level of violence in some of these wars have increased. This was not the case during the Cold War, when the ideological, political, military and economic structures of bipolarity overshadowed almost all of these intra-state wars, relegating them to a peripheral irrelevance.

Where *The Utility of Force* – and its main argument about a paradigm shift from conventional war to 'war amongst the people' – presents the contemporary character of war erroneously is the notion that post-Cold War Western military experiences should influence thinking about the utility of force in the future. Western wars of the post-Cold War era have mostly been costly mistakes with non-existent strategic interests at stake. In most cases, they are wars that should never have been waged. Afghanistan after early 2002, the war against Iraq (formally 2003–2011, but in practice continuing in some form even today) and Operation Unified Protector in Libya (2011) are prime examples of the flawed Western strategic decisions to go to war in order to achieve the unachievable by force of arms.

This chapter argues that 'war amongst the people' is one perspective on warfare amongst many others that have been developed to cope with the problem that the short-sighted post-Cold War era Western security analysis had identified: the need to settle 'new' kinds of wars with Western militaries out-of-area. 'War amongst the people' identifies mostly tactical and operational level lessons identified from post-Cold War era conflicts that have served limited strategic purpose for the West. This was made possible by the threat vacuum caused by the demise of the Soviet Union. With that threat gone, Western statesmen and analysts of strategic affairs became troubled by the loss of solid foundations for planning and executing security and defence policy, including formulating guidelines for long-term military capability development. Unfortunately, 'war amongst the people' did not address the strategic level – analysing *why* Western states should manage intra-state wars around the world with their militaries. Militaries are suitable tools of statecraft for deterrence and the use of large-scale high-quantity and high-quality violence. Doing – or trying to do – nation-building, humanitarian missions or stability operations is to use force in a way that is incompatible with war's nature. The use of large-scale violence is good at destroying targets, but will rarely be able to solve complex social problems with ethnic, religious, tribal or other motivations.

---

2 For the purposes of this discussion, the terms 'war' and 'conflict' are used interchangeably.

This chapter proceeds in three stages. First, the framework is constructed by analysing the end of the Cold War and the way that Western states have redefined international security and war during the post-Cold War era. In the second stage, the 'war amongst the people' thesis is scrutinised as one expression of many similar attempts to formulate a coherent analytical construct to come to terms with the changing character of war. In the third and final stage, the consequences of changed Western perspectives on war are analysed.

The Root Cause

It is well-known that the number of armed conflicts started to increase after the end of the Second World War. The rising trend continued until the very end of the Cold War. Most – a vast majority – of these armed conflicts were 'below' the state level – intra-state wars (civil wars) and extra-state conflicts related to the demise of colonial empires.[3] What was remarkable throughout the Cold War was the almost non-existent interest in wars below the state level. As the main parties to the Cold War were focused on the ideological confrontation and superpower rivalries, dozens of ongoing armed conflicts in the 'periphery' were almost completely neglected or gained very limited attention, unless, like the conflicts in Vietnam and Afghanistan, they were directly related to the waging of the Cold War.

The material aspects of the Cold War were manifested by large-scale conventional armed forces of both sides containing dozens of army corps, divisions, brigades and other military units, totalling millions of soldiers. In addition, thousands of nuclear warheads were produced and stockpiled to serve the purpose of deterrence – and possibly even a potential war. In addition to this material dimension, a 'Cold War of the mind'[4] also developed during the threat-penetrated decades of the confrontation. With the constant possibility of large-scale war breaking out, the main protagonists of the Cold War came to terms with the rules of the Cold War game, developing doctrines and capabilities for offsetting the opponent's potential advantages and to ameliorate their own relative standing vis-à-vis their opponents. Thus, between 1945–1989, a set of rules concerning

---

3   See Scott Gates, Håvard Nygård, Håvard Strand and Henrik Urdal, *Trends in Armed Conflict 1946–2014: Conflict Trends 01/2016* (Oslo: Peace Research Institute, 2016): available at https://www.prio.org/utility/DownloadFile.ashx?id=8&type=publicationfile, accessed 10 October 2018.

4   Ken Booth, 'Cold Wars of the Mind' in Ken Booth (ed.), *Statecraft and Security: The Cold War and Beyond* (Cambridge: Cambridge University Press, 1998), 29–55.

international security and defence developed, matured – thereby facilitating the Cold War era adversaries' engagement with each other – and ultimately ossified.

In contrast, the immediate post-Cold War era was a generally acknowledged time of transition, although the end point of this process of transition was hard to discern. As the 1990 *Charter of Paris for a New Europe* declared, 'The era of confrontation and division of Europe has ended. We declare that henceforth our relations will be founded on respect and co-operation. Europe is liberating itself from the legacy of the past'.[5] As decades of bipolar confrontation came to an abrupt end, the West rapidly lost longstanding foundations concerning the rules of the international security game and the associated principles to maintain and develop national armed forces. In an instant, the rationale for maintaining large-scale mechanised forces seemed anachronistic. Similarly, the North Atlantic Treaty Organisation (NATO) was perceived to be rapidly losing its *raison d'être* as deterrence against the Warsaw Pact or the Soviet Union became almost meaningless with the winding down of the superpower confrontation during the early 1990s. By 1992, both the Warsaw Pact and the Soviet Union had disintegrated and Russia – the successor state of the Soviet Union – started to deal with enormous domestic problems concerning social cohesion, economic difficulties and the threat of territorial and political disintegration. The future survival of Russia was uncertain, which meant that it posed practically no military threat to the West.

In the US, the 1994 *Department of Defense Annual Report* to the President and Congress explicitly stated the 'problem':

> This is a period comparable to the end of World War II. It was clear that profound change had taken place, but it was unclear what kind of world would replace the old one. Today, it is not clear what new paradigm will replace East-West rivalry and a bipolar world, but one can see clear threats to America and its interests … Defining the post-Soviet security environment is the critical first step in sizing and shaping a new defense, right for the times.[6]

The end of the Cold War was truly a conceptual watershed event that had far-reaching consequences for the way that Western states conceptualised international security and the role of their armed forces in the future. Herein lay the problem: the sudden and unexpected demise of the Soviet Union caused a vacuum in Western threat perceptions. Suddenly – in a

---

5   The text of the 1990 *Charter of Paris for a New Europe* is available at https://www.osce.org/mc/39516?download=true, accessed 10 October 2018.

6   See Department of Defense, *Annual Report to the President and the Congress by Secretary of Defense Les Aspin* (Washington DC: US Government Printing Office, January 1994), 1.

matter of years – the defining military threat to the West and its military institution for collective defence – NATO – was gone.[7] Moreover, in the short term, nothing new appeared to replace the rapidly lowering threat of massive military attack on Central and Western Europe. Simply stated, the West had won the Cold War, but, as Western statesmen were celebrating their victory, they were also having their *Was nun*-moment: what kind of new international rules would surpass the old threat that had provided the rationale for the maintenance and development of Western armed forces for decades and which had glued the member states of NATO together?[8]

This period was marked by many attempts to come to grips with the emerging reality of the post-Cold War era. In 1991 George H.W. Bush heralded the emergence of a 'New World Order'.[9] Two years previously, Francis Fukuyama published his seminal article noting 'An End of History?', in which he argued that it was 'not just the end of the Cold War, or the passing of a particular period of post-war history, but the end of history as such: that is, the end point of mankind's ideological evolution and the universalization of Western liberal democracy as the final form of human government'.[10] Three years later, Fukuyama developed the essay into an influential book. One of the visions of the post-Cold War era, his *End of History* became a must-read for scholars and practitioners of International Relations.

In addition to the above-mentioned attempts to redefine and redesign – or at least influence – the post-Cold War era logic of international security, many others started to accumulate. The emergence of 'globalisation';[11] 'the unipolar moment';[12] 'an agenda for peace';[13] the appearance of 'new wars',[14]

---

7   Anna Wieslander, 'NATO, the U.S. and Baltic Sea Security', *UI Paper* 3 (Stockholm: Swedish Institute of International Affairs, 2016) available at https://www.ui.se/globalassets/butiken/ui-paper/2016/nato-the-u.s.-and-the-baltic-sea-security---aw.pdf, accessed 10 October 2018.

8   Jyri Raitasalo, 'Moving beyond the Western "Expeditionary Frenzy"', *Comparative Strategy* 33:4 (2014), 372–88.

9   George H.W. Bush, *Address before a Joint Session of the Congress on the Persian Gulf Crisis and the Federal Budget Deficit* (Washington DC: US Government Printing Office, 1990), available at http://www.presidency.ucsb.edu/ws/?pid=18820, accessed 10 October 2018.

10 Francis Fukuyama, '"The End of History?", *The National Interest* 16 (1989), 1–18.

11 Roland Robertson, 'The Globalisation Paradigm: Thinking Globally' in D. Bromley (ed.), *Religion and Social Order* (Greenwich: JAI Press, 1991), 207–24.

12 Charles Krauthammer, 'The Unipolar Moment', *Foreign Affairs* 70:1 (1990/1991), 23–33.

13 United Nations, *An Agenda for Peace: Report of the Secretary-General pursuant to the statement adopted by the Summit Meeting of the Security Council on 31 January 1992* (New York: United Nations, 1992), available at http://www.un-documents.net/a47-277.htm, accessed 10 October 2018.

14 Mary Kaldor, *New and Old Wars: Organized Violence in a Global Era* (Cambridge: Polity Press, 1999).

'non-trinitarian wars'[15] and 'low intensity conflicts'[16] as well as 'the clash of civilizations'[17] were all part of Western attempts to conceptualise the emerging security environment and the challenge that it would pose to the West. What these concepts have in common is that they all purport to describe or define some important aspects of the post-Cold War world at the level of strategy or security policy.

In the military domain specifically – oscillating between military strategic, operational and tactical perspectives – many similar analytical constructs emerged after the Cold War had ended. The aim of the new constructs was to shed light on the emerging realities of the post-Cold War era in general and to provide guidelines for the development of armed forces in the future in particular. These new concepts included 'Revolution in Military Affairs', 'Peace Operations', 'Military Crisis Management', 'Crisis Response Operations', 'Military Operations Other Than War', 'Humanitarian Interventions' or 'Three Block War'.[18] This proliferation of prediction provides the wider context into which Smith's 'war amongst the people' thesis was articulated.

Thus, during the 1990s and the following decade, sedimented notions of war – and especially of future war – underwent significant transformation. As has been noted already, there were many attempts to redefine the emerging international security logic and the role that armed forces would play in the future, of which Smith's was only one. These new concepts were in high demand as statesmen sought suitable ways to redirect their perspectives on security and the guiding principles underpinning the development of future armed forces in accordance with the emerging – and potentially new – rules of the international security game.

Redefining War

The process of redefining shared Western notions on war and warfare began with the attempt by Saddam Hussein to boost his power position by occupying Kuwait in August 1990. Coinciding with the end of the Cold War, the 1991 Gulf War gained extra significance: purportedly showing

---

15 Martin van Creveld, *The Transformation of War* (New York: Free Press, 1991).

16 For details, see US Armed Forces, *Military Operations in Low Intensity Conflict FM 100-20/ AFP 3-20* (Washington DC: Department of Defense, 1990), 1–1.

17 Samuel P. Huntington, 'The Clash of Civilizations?', *Foreign Affairs* 72:3 (1993), 22–49.

18 See Jyri Raitasalo, *Constructing War and Military Power after the Cold War – The Role of the United States in the Shared Western Understandings of War and Military Power in the Post-Cold War Era* (Helsinki: National Defence College, 2006), available at https://www.doria.fi/bitstream/handle/10024/74436/StratL1_21.pdf?sequence=1&isAllowed=y, accessed 10 October 2018.

the future path of military conflicts as the Soviet Union and the West were transitioning from the world of the Cold War towards less hostile relations and potentially even a new world order. In reality, if there was one lesson learned from the Gulf War, it was that modern high-tech military capabilities proved highly effective against traditional lower-tech massed conventional armed forces. In 1992 it was noted that there was a Military-Technical Revolution ongoing and that its effects proved decisive in the war the previous year.[19] This lesson was drawn not only in the US, but in Russia and China as well, although both Russia and China were hindered by their relative military deficiencies.

As the decade progressed, the US embarked on substantive military transformation. Conflicts in Bosnia (1995) and Kosovo (1999) – as well as the almost constant use of air power against Saddam Hussein's regime in Iraq throughout the 1990s – provided empirical evidence to support cutting military mass and replacing it with new high-tech network-centric capabilities. By 1999, it was becoming obvious that a capability gap was developing across the Atlantic Ocean, as most European states lacked the willingness and the resources to follow suit and procure their armed forces with the most expensive – but very effective – networked military systems. After all, there was no significant military threat in Europe. Nevertheless, European states tried to follow the US lead as the 1999 NATO Defence Capabilities Initiative – and the 2002 Prague Capabilities Commitment – highlighted.[20] Both projects were intended to keep the Europeans on board with developments in the US.

The increasing precision, force protection, stealth and C4ISR capabilities that the RMA proponents were arguing for was one fact that facilitated the increasing military activity of the West – under US leadership – to engage in the so-called 'new wars', ethnic conflicts or non-trinitarian wars that were ongoing in the post-Cold War international system. In fact, the RMA thesis created a 'push' for the Western states to intervene in 'new' kinds of conflict out-of-area in order to stay militarily relevant in the post-Cold War era. Although the number and deadliness of armed conflicts in the world started to decline in the very early 1990s, more attention was placed on them after the mental straitjacket of the Cold War era bipolar confrontation was over. As the 2005 *Human Security Report* noted:

---

19 See the section on 'Conduct of the Persian Gulf War' in Department of Defense, *Final Report to Congress Pursuant to Title V of the Persian Gulf Conflict Supplementation Authorization and Personnel Benefits Act of 1991* (Washington DC: Department of Defense, 1992).

20 NATO, *Defence Capabilities Initiative, NATO Press Release NAC-S(99)69* (Brussels: NATO, 1999); NATO, *Prague Summit Declaration – Issued by the Heads of State and Government Participating in the Meeting of the North Atlantic Council in Prague on 21 November 2002: NATO Press Release 127* (Brussels: NATO, 2002).

Since the end of the Cold War, there has been a dramatic and sustained decline in the number of armed conflicts. And an uneven but equally dramatic decline in battle-deaths has been under way for more than half a century. Yet these facts remain largely unknown, in part because there are no reliable, official global statistics.[21]

Despite this decline in the number of armed conflicts and related battle-deaths, Western states remained fixated on managing them. The notion of an ongoing RMA has provided a way to intervene – mostly by air power – in places where military conflicts made headlines and humanitarian suffering called for something to be done in such theatres as Iraq, Somalia, Haiti, Bosnia, Kosovo, East-Timor, Liberia and Sierra Leone.

Shedding the additional bulk of the Cold War era and focusing on new high-tech RMA-capabilities, many European states moved from large-scale armed forces based on big reserves to all-volunteer forces. When the Cold War ended, four European states – Ireland, Luxembourg, Malta and the UK – had all-volunteer professional military forces. Beginning in the early 1990s, most others abandoned conscription and the mass mobilisation model.[22] Arguably, all-volunteer forces are easier to send to missions – to fight a war – than a conscripted force of warrior-citizens, at least when the mission is not directly related to the security of the state, as may be the case with 'wars amongst the people'. This trend of loosening constraints on the use of military force in expeditionary operations has been amplified by the greater privatisation of war, with contracting tasks traditionally taken care of by the militaries themselves undertaken by private military companies (PMCs).

Relying on contractors facilitates keeping peace-time military organisations 'lean', because, in a time of crisis, it is possible to resort to the 'surge capability' provided by different contractors, for example in the fields of logistics, maintenance, providing security and other supporting tasks. In addition, losing a contractor in an operation is more politically palatable than losing a professional soldier. While the trend of shifting supporting military missions to the private sector has been under way at least for two decades, it has been the US-led campaigns in Afghanistan and Iraq that have made public the increased scope and domain of this process of privatising war.[23]

---

21 University of British Columbia Human Security Centre, *War and Peace in the 21st Century: Human Security Report 2005* (Oxford: Oxford University Press, 2005), 17.

22 Bernard Boene, 'Shifting to All-Volunteer Armed Forces in Europe: Why, How, With What Effects?', *Forum Sociologico* 19 (2009), 1.

23 Jyri Raitasalo, 'Reconstructing Finnish Defence in the Post-Cold War Era', *Finnish Defence Studies* 18 (Helsinki: National Defence University, 2010), 30–32.

After the 1990s, the threshold on the use of military force within the international system was lower than it was during the Cold War era. This was due to Western military interventions in the name of military crisis management or humanitarian interventions to stop humanitarian catastrophes and facilitate statebuilding in conflict areas. After the terrorist attacks in the US on 11 September 2001 (9/11), the political attractiveness of a military response to counter the 'new' threat of terrorism was near irresistible. Thus, the way was paved for a highly militarised response: the GWOT.

The GWOT started a long sequence of events and unintended consequences that have defined much of the Western approach to international security in the post-9/11 era. Focusing on terrorist organisations and even individual terrorists globally has further expanded the scope of Western military activity. Relying on the out-of-area expeditionary experiences of the 1990s, during the next decade many political inhibitions to the use of military force were removed and the rules for the use of military force by the US and many other Western states were redefined. Unilateral use of military force in a preventive fashion was espoused as a necessity, together with the ideology-infused concept of democracy promotion.

The two-term George W. Bush Administration (2001–2009) redefined the American – and arguably the 'Western' – perspective on war. The consequences of this redefinition are potent and enduring. The havoc caused by the 2003 invasion of Iraq – with its unintended consequences – together with NATO's military intervention in Libya, should necessitate some form of Western response for years to come. Unfortunately, the serious challenge posed by a resurgent Russia and a rising China is going to demand more focus by the West, which seems to have ended up with a 'two front war'. This is where the 'war amongst the people' thesis has a role to play as it is predicated on the notion that the West *should* get involved militarily in violent crises out-of-area.

Getting the Story Wrong – at the Strategic Level

The first analytical part of Smith's thesis is easy to agree with. Based on the fact that practically none of the wars since the Second World War have been between states – and shaped by personal experiences from several Western military operations – he argues convincingly that state-on-state wars are extremely rare. For decades, most wars have included or been fought by non-state actors.[24] Of course, this argument is not particularly novel. Writing some 14 years previously, Martin van Creveld noted that:

---

24   Gates, Nygård, Strand and Urdal, *Trends in Armed Conflict.*

We are entering an era, not of peaceful economic competition between trading blocs, but of warfare between ethnic and religious groups. Even as familiar forms of armed conflict are sinking into the dustbin of the past, radically new ones are raising their heads ready to take their place ... conventional military organizations of the principal powers are hardly even relevant to the predominant form of contemporary war.[25]

Smith's thesis goes beyond the non-state war argument and proposes that the *aims* of war have changed in a Kuhnian style paradigm shift from industrial war towards 'war amongst the people'. Decisive victories are a thing of the past as wars are fought to create a condition where strategic aims can be achieved using other means. According to the argument, the destruction of the enemy is simply not enough. War has become more about altering the opponent's intentions. War is also about perceptions – the minds of people. In order to succeed in this new kind of war, the use of military force must be part of a more comprehensive set of capabilities and tools that serve politically defined goals. The Cold War era military 'machines' and doctrines would be ill-suited to a world where 'war amongst the people' is the dominant paradigm.

The problem with the 'war amongst the people' construct is located behind the description of armed conflicts in the world after the Second World War. What Smith fails to show is critical: even if most wars are not based on what he calls the industrial war paradigm why should any of the Western states be interested in dealing with the proliferation of these 'new' wars? What is the strategic rationale for the active use of military force in a world with limited direct conventional military threats to the West? Following the demise of the Cold War, there have been between 30–50 wars annually within the international system.[26] Yet the 'war amongst the people' thesis fails to show the relevance and utility of the West interfering in these wars.

Thus, the thesis is based on the conflation of those wars that are the most commonplace in the contemporary international system with those that really matter for the West. This is not a deficiency of the 'war amongst the people' thesis alone. In search of 'new' post-Cold War era guidelines for security and defence, many Western statesmen felt provoked by events to take a stand – to do something – regarding so-called 'new wars' or 'wars amongst the people' – in Europe, the Middle-East, Africa, Asia and even in the Western Hemisphere, although in reality their actions did not have to be military in nature. Realistically, nothing forced the West to

---

25 Van Creveld, *Transformation of War*, 20.
26 Van Creveld, *Transformation of War*, 20.

take a stand militarily, beyond their own rhetoric that something must be done. Based on incremental decision-making over time, the West drifted towards responding and reacting militarily to complex wars on several continents. This happened without a clear-cut vision or a plan to move from the deterrence and large-scale warfighting tradition towards military crisis management, humanitarian missions, statebuilding and democracy promotion.[27]

One of the key facilitating factors boosting constructs like 'war amongst the people' within the Western security community has been the notion of globalisation and its effects on the ground rules between states. This globalisation narrative has accentuated the increasing interdependence between states and the fact that borders between states have become ever more porous. In addition, and more relevant to the defence realm, the globalisation narrative has suggested that the best way to provide security in this 'new' epoch is cooperative security, based on a non-zero-sum approach to politics. The problem with the notions of cooperative security and the non-zero-sum game is related to the fact that not all relevant state actors share this optimistic vision of collectively providing security by managing the globalising international system: co-operation will only work if all are prepared to do so. Even though many in the West thought – or hoped – that the end of the Cold War meant the end to spheres of influence, power politics and great power rivalries, more recent events have proved that it did not. Russia and China have not accepted the Western globalisation narrative, nor have they acted according to its logic. As early as 1999, when NATO was preparing to start a war against Serbia under the auspices of a humanitarian intervention, Russian President Boris Yeltsin threatened the West with war, noting that 'I told NATO, the Americans, the Germans, don't push us towards military action. Otherwise there will be a European war for sure – and possibly world war'.[28] The list of signs that non-Western great powers have not accepted Western interpretations of globalisation goes on: President Vladimir Putin's 2007 Munich Security Conference speech blaming the West (read the US) for hyper-active use of military force,[29] the war between Russia and Georgia in 2008, the increased assertiveness of China in the South and East China Seas, the annexation of the Crimean peninsula by Russia in 2014, Russia's proxy war in Eastern Ukraine since 2014 and Russia's influence operations in Syria since 2015 are

---

27 Raitasalo, *Expeditionary Frenzy*.

28 'Kosovo "could spark world war"', *BBC News* 9 April 1999, available at http://news.bbc. co.uk/2/hi/europe/315220.stm, accessed 10 October 2018.

29 Vladimir Putin, *Speech at the Munich Conference on Security Policy*, available at http:// en.kremlin.ru/events/president/transcripts/24034, accessed 10 October 2018.

all clear challenges to the post-Cold War Western notions of globalisation and cooperative security. Given this return to conventional great power activity – and potential competition – 'war amongst the people' may prove ultimately to be passé.

The Long-term Consequences

Changing focus from deterrence, as well as large-scale mechanised high-quality and high-quantity violence against advanced military great powers, to facilitating nation-building, humanitarian objectives and counterinsurgency operations was an almost twenty-year process during which Western states tried – not always successfully – to grasp the post-Cold War era new rules of international security and the nature of contemporary war. The articulation of the 'war amongst the people' thesis was part of this process. Such constructs have influenced Western strategy formulation, defence planning processes and threat perceptions, whether necessary or not. One core aspect of defence planning is related to the definition of military threats to national security. Thus, it is in the process of defence planning where core military tasks for the armed forces are defined.

The perspective of defence planning and military capability development looks 10–20 years into the future. It is a process that is future-oriented, but one that has to take into account the legacy of the pre-existing military procedures, doctrine and platforms. All national armed forces have been constructed incrementally – building new layers onto the preceding ones. In the defence realm, the shadow of history is cast way into the future. This has been problematic, because the vision of future war that 'war amongst the people' proposes is so different from the pre-existing legacy forces and doctrines. In effect, the thesis proposed a 'U-turn' in defence planning and military capability development, without giving sufficient consideration to the likely return of history. The problem with U-turns in the field of defence is that they are difficult to execute quickly. A 'quartal' in defence planning is 25 years. Trying to make a quick U-turn can only lead to loss of capability in the short term (5–10 years) as developing, procuring and training new capabilities takes many years. From a longer-term perspective, something between 10–30 years, these U-turns may produce new kinds of tools, doctrines, organisations and military ethos.

However, if the U-turn taken was based on faulty reading of the (future) security environment and the strategic interests of the state in question, it can lead to tragic consequences: long-term loss of military capability, effectiveness and credibility. The argument against the 'war amongst the people' thesis is based precisely on this factor. It proposed strategic level changes for defence planning as well as security and defence policies that are based on empirical (real life) tactical and operational level lessons identified or lessons learned from past and ongoing military operations. Just because some recent-past Western military operations have failed – or they have not succeeded – does not mean that states should engage in similar kinds of operations in the future – even with new doctrines, organisations and military equipment. Past failures should be a guide into the future. Learning from one's mistakes at the strategic level would entail lessening one's willingness to use military force in crises that have little or no significance for one's vital security interests. In layman's terms, this means getting rid of the strategic trigger-happiness that has gradually sneaked into Western security and defence policy during the post-Cold War era.

Defence policy cannot be formulated in the trenches. Nor can states' strategic interests be defined through the prism of succeeding in wars that are the most commonplace in the world. Democracy promotion, dictator-decapitation and remaking the world through the force of arms have not produced successes during the post-Cold War era. Afghanistan, Iraq and Libya are some recent reminders of that. So far, the post-Cold War era Western military social engineering has back-fired. And this has happened for a good reason: military force has been used too frequently in missions that have no sensible military solution. Even with the best of intentions, the West has been – and continues to be – incapable of solving complex violent political crises around the globe with existing or potentially even future military tools. Trying to change this – relying on the 'war amongst the people' thesis or some other similar construct – would only perversely increase the security threats against the West and institutionalise a myopic strategic perspective that facilitates picking the wrong battles, diverting attention from other more serious matters within the more traditional, conventional realm.

Conclusion

Rupert Smith's notion of the lowered utility of force in many contemporary conflicts is accurate: military force has no utility in many of the political and social problems of the world. However, responding to this by clinging to constructs like 'war amongst the people' is neither helpful nor viable and would severely threaten Western security by continuing to embroil states in political and social crises that have little or no connection to their strategic security interests. Furthermore, accepting 'war amongst the people' as a guideline for the development of military forces would lead to further atrophying of large-scale deterrence and warfighting capability within the West in general and Europe particularly, at a time when this military capability will more likely be needed during the decades to come. The post-Cold War era is over – and so is America's unipolar moment. Great power politics is back, and it is back with a vengeance. Hyped concepts within the defence realm usually have a short lifecycle. Ultimately – and with some relief – this also applies to 'war amongst the people'. With the advent of a multipolar world order and great power rivalries, it is time to note that 'war amongst the people' had its time and its place, but now needs to be consigned to the dustbin of history. Instead, the West needs to start preparing properly for a future defined by zero-sum great power competition. It is time to bring deterrence and real defence back to a more prominent position on the Western military agenda.

References

Boene, B., 'Shifting to All-Volunteer Armed Forces in Europe: Why, How, With What Effects?', *Forum Sociologico* 19 (2009).

Booth, K., 'Cold Wars of the Mind' in Ken Booth (ed.), *Statecraft and Security: The Cold War and Beyond* (Cambridge: Cambridge University Press, 1998).

George H.W. Bush, *Address before a Joint Session of the Congress on the Persian Gulf Crisis and the Federal Budget Deficit* (Washington DC: US Government Printing Office, 1990), available at http://www.presidency.ucsb.edu/ws/?pid=18820, accessed 10 October 2018.

Department of Defense, *Final Report to Congress Pursuant to Title V of the Persian Gulf Conflict Supplementation Authorization and Personnel Benefits Act of 1991* (Washington DC: Department of Defense, 1992).

Department of Defense, *Annual Report to the President and the Congress by Secretary of Defense Les Aspin* (Washington DC: US Government Printing Office, January 1994).

Fukuyama, F., 'The End of History?', *The National Interest* 16 (1989).

Gates, S., Nygård, H., Strand, H. and Urdal, H., *Trends in Armed Conflict 1946–2014: Conflict Trends 01/2016* (Oslo: Peace Research Institute, 2016).

Gray, C.S., 'How Has War Changed Since the End of the Cold War?', *Parameters* XXXV:1 (2005).

Huntington, S.P., 'The Clash of Civilizations?', *Foreign Affairs* 72:3 (1993).

Kaldor, M., *New and Old Wars: Organized Violence in a Global Era* (Cambridge: Polity Press, 1999).

Krauthammer, C., 'The Unipolar Moment', *Foreign Affairs* 70:1 (1990/1991).

NATO, *Defence Capabilities Initiative, NATO Press Release NAC-S(99)69* (Brussels: NATO, 1999).

NATO, *Prague Summit Declaration – Issued by the Heads of State and Government Participating in the Meeting of the North Atlantic Council in Prague on 21 November 2002: NATO Press Release 127* (Brussels: NATO, 2002).

Putin, V. *Speech at the Munich Conference on Security Policy,* available at http://en.kremlin.ru/events/president/transcripts/24034, accessed 10 October 2018.

Raitasalo, J., *Constructing War and Military Power after the Cold War – The Role of the United States in the Shared Western Understandings of War and Military Power in the Post-Cold War Era* (Helsinki: National Defence College, 2006).

Raitasalo, J., 'Reconstructing Finnish Defence in the Post-Cold War Era', *Finnish Defence Studies* 18 (Helsinki: National Defence University, 2010).

Raitasalo, J., 'Moving beyond the Western "Expeditionary Frenzy"', *Comparative Strategy* 33:4 (2014).

Robertson, R., 'The Globalisation Paradigm: Thinking Globally' in D. Bromley (ed.), *Religion and Social Order* (Greenwich: JAI Press, 1991).

Smith, R., *The Utility of Force: The Art of War in the Modern World* (New York: Knopf, 2005).

United Nations, *An Agenda for Peace: Report of the Secretary-General pursuant to the statement adopted by the Summit Meeting of the Security Council on 31 January 1992* (New York: United Nations, 1992), available at http://www.un-documents.net/a47-277.htm, accessed 10 October 2018.

University of British Columbia Human Security Centre, *War and Peace in the 21st Century: Human Security Report 2005* (Oxford: Oxford University Press, 2005).

US Armed Forces, *Military Operations in Low Intensity Conflict FM 100-20/AFP 3-20* (Washington DC: Department of Defense, December 1990).

Van Creveld, M., *The Transformation of War* (New York: Free Press, 1991).

Wieslander, A., 'NATO, the U.S. and Baltic Sea Security', *UI Paper* 3 (Stockholm: Swedish Institute of International Affairs, 2016).

3

MANAGING 'WAR AMONGST PEOPLES'

Alex Waterman

Introduction

Our understanding of 'war amongst the people' is rooted in the theories of insurgency and counterinsurgency that emerged from late colonial and early Cold War rebellions. Mao Tse Tung's insurgents conceptualised their relationship to the population as 'fish among water'.[1] To counter this strategy, counterinsurgents subsequently modelled their doctrinal principles on how to remove this fish from the water. However, the analogy overlooks the complex ecosystem that the 'fish' inhabits. Certain conditions for survival, such as atmospheric pressure, depth, salt levels and its place in the food chain are not constants but differ over space and time. For the fisherman, some parts of the sea are more accessible than others, requiring a trade-off between costs, return and willpower, pointing towards a degree of complexity in the theorisation of insurgencies that Mao's original analogy lacks.

The understanding of 'the people' in insurgency and COIN studies assumes the ability of either party to 'win' this overall entity to achieve critical mass. Consequently, 'the people' are theoretically located in an almost passive position between revolutionary and counter-revolutionary in a two-way, somewhat linear contest. The concept of 'population-centric COIN' as a method to counter this has endured well beyond the Cold War and continues to feature in contemporary Western military doctrines, producing a set of benchmarks for measuring success and failure.

---

1 Mao Tse-Tung, *On Guerrilla Warfare* (Washington DC: Presidio Press, 1991), 189, available at http://www.marines.mil/Portals/59/Publications/FMFRP%2012-18%20%20Mao%20 Tse-tung%20on%20Guerrilla%20Warfare.pdf, accessed 16 October 2018.

However, as practitioners learned in COIN operations in Afghanistan and Iraq, interactions with 'the people' constitute interventions into complex, fragile and evolving relations between the diverse 'peoples' that make up the socio-political environment. Such peoples may vary considerably in both scale and the degree of influence that they exert in a conflict environment, ranging from large ethno-political formations and economically-influential social classes to much more micro-level social formations, such as factional or clan groupings within village politics. Winning over one community or interest group may, in some circumstances, antagonise another or disrupt relationships established with other constituencies. Furthermore, counterinsurgents may lack the political and logistical resources to 'win' certain segments of the people. For some political leaders, certain 'peoples', given their ability to affect wider power calculations, may be more important or indeed more amenable to 'purchase' than others. Indeed, the acceptable reality for some, albeit not all, counterinsurgents of winning enough 'peoples' to establish an equilibrium that ensures regime survival and fulfils core political goals also produces stark variations in the methods adopted to prosecute COIN campaigns that are often overlooked in the accompanying literature.

This chapter therefore contends that the notion of the 'war amongst the people', rooted in out-dated ideological dichotomies associated with Maoist insurgency theory, is conceptually insufficient for unpacking these complex dynamics. It calls for an analytical refocusing towards 'war amongst peoples' to account for the complexity and dynamism of counterinsurgent-societal interactions in complex insurgency environments. The Syrian Civil War is offered as a salient case in point that presents a challenging set of circumstances for the incumbent counterinsurgent. A plethora of insurgent groups base their support on political claims to represent numerous ethnic, religious and community-based interest groups. The Syrian government lacks the political, economic and military resources to 'win' over 'the people', as understood in the Maoist model. Instead, Damascus has sought to win over 'peoples', in doing so cutting into and across ethnic lines to disrupt the consolidation of power against it, while simultaneously employing polarising, sectarian strategies in a bid to retain the regime's minority constituencies and deter possible defections.[2] These efforts to engage with, 'buy off' and cultivate relationships with particular 'peoples' or interest groups demonstrate the multifaceted nature of state–population interactions, whilst highlighting the oft-overlooked socio-political nuances behind a conflict characterised

---

2   Raymond Hinnebusch, 'Syria' in Ellen Lust (ed.), *The Middle East* (Thousand Oaks: CQ Press, 2017), 799.

by brutal violence. Outlining this conceptual framework presents opportunities for further research assessing the relationship between existing state attempts to analyse and negotiate these complex socio-political ecosystems and the broader principles, doctrines and concepts underpinning contemporary COIN theories.

Accordingly, the chapter begins by reviewing current theorisations of 'the people' in classical and contemporary insurgency and counterinsurgency theory. It argues that 'the people' is essentially rooted in the strategy of the Maoist theoretical tradition of insurgencies. It then introduces the notion of 'war amongst peoples', contending that this framework encapsulates a broader range of counterinsurgent-population interactions than notions of 'the people' and thus offers an analytically more useful point of departure for analysing counterinsurgent-population interactions. Thereafter, the chapter contextualises and provides an overview of the Syrian state's fundamentally political engagement with constituencies crucial to regime survival both prior to and during the Syrian Civil War, arguing that 'war amongst peoples' captures a broader range of methods by which states seek to engage in this environment. It concludes by suggesting that theories of insurgency and COIN need to better account for these political considerations and their impact on how counterinsurgents engage with socio-political constituencies, both in authoritarian COIN campaigns in states such as Syria, but also in advanced democracies conducting COIN campaigns amidst 'wars amongst peoples'.

People or Peoples?

Rupert Smith first made the case for the emergence of 'war amongst the people' as a paradigm with which to understand much of contemporary conflict, basing this on six key observations that essentially juxtapose this new paradigm with the large-scale, decisive absolutes of Industrial Interstate War.[3] In this form of war, Smith argued, 'the strategic objective is to capture the will of the people and their leaders, and thereby win the trial of strength'.[4]

---

3   For an overview of these six trends, see Rupert Smith, *The Utility of Force: The Art of War in the Modern World* (London: Penguin, 2006), 17. This chapter does not engage with Smith's argument concerning the transformational nature of the 'war amongst the people' paradigm, which has been engaged with elsewhere. See Alexander McKenzie, '"New Wars" Fought "Amongst the People": "Transformed" by Old Realities?', *Defence Studies* 11:4 (2011), 569–93.

4   Smith, *The Utility of Force*, 277.

Maoist Theory and 'The People'

Regardless of whether or not 'war amongst the people' constitutes a novel paradigm, the centrality of the people in revolutionary and counter-revolutionary warfare is not new. Indeed, colonial 'small war' practitioners acknowledged the importance of local populations in colonial policing operations, although conceptualisations of and approaches to interactions with the population varied significantly.[5] However, the idea that the population can be harnessed to enact political change is embedded within social revolutionary theory. This is particularly the case within Marxist ideology, which conceptualises the proletarian masses as the vehicle for political change, provided that they are mobilised into a state of revolutionary consciousness. In other words, strategic success is possible if the will of the people is captured. These ideas were moulded into guerrilla strategy during the Chinese Civil War by Mao, consolidating the role of 'the people' as a core political element of insurgency theory. Following their retreat into the mountains in 1927, Mao referred to the three-phased plan adopted by the Chinese Communist Party (CCP) to ideologically emancipate the Chinese people, liberate the population from Japanese occupation and, by doing this, capture political power.[6]

The first phase, consisting of education, organisation, preparation and mobilisation sought to convince 'the people' of the class struggle. The second phase sought to secure the consolidation and expansion of the 'liberated zones' in which 'the people' had committed their political support, while the third marked the point of critical mass from which the decisive offensive could be launched.[7] In this model, the understanding of 'the people' was intrinsically linked to the party's ideological dispensation towards securing revolution using the majority rural Chinese population, identified through the political-ideological prisms of class struggle and national liberation. Mao thus conceived of the guerrilla as acting 'in the interests of the whole people or the greater part of them', with a 'broad basis in the national manpower'. For him, an insurgency that contradicts the interests of this whole was 'easy to destroy because they lack a broad

---

5   For example, while discussing punitive operations, C.E. Callwell suggested that care should be taken to avoid excessive measures against villages that were 'merely victims'. See Charles Edward Callwell, *Small Wars: Their Principles and Practice* (London: His Majesty's Stationery Office, 1906). C. Gwynn on the other hand referred to hostiles as 'fellow citizens' of the empire that would, at some stage, be reintegrated into the daily life of colonial civil governance, therefore advocating the employment of minimum force. See Charles W. Gwynn, *Imperial Policing* (London: Macmillan, 1934), 14.

6   Mao, *On Guerrilla Warfare*, 42–43.

7   Ian Beckett, *Modern Insurgencies and Counter-Insurgencies: Guerrillas and Their Opponents since 1750* (New York: Routledge, 2001), 74–75.

foundation in the people'.[8] This provides the foundations for the symbiotic analytical relationship Mao establishes between the population and the insurgent, with shared interests and mutual dependency, making the population the water within which the fish swims. For Mao, insurgents must therefore seek to demonstrate perceived shared interests with the population by adopting techniques to garner favour amongst them. In the Chinese context, these included politeness, the replacement of used goods, assistance with sanitation and welfare and the dissemination of propaganda using traditional methods, such as plays and poems.[9] This was buttressed by other crucial strategies of control, such as a pervasive organisational and coercive apparatus, which, while critical, has received much less attention in the wider COIN and insurgency literature, perhaps as part of the desire to demonstrate the 'moral superiority of the guerrilla' or indeed his opponent.[10]

Following the CCP's success in China in 1949, which coincided with the beginnings of decolonisation and the emergence of the Cold War, a number of anti-colonial insurgencies, including those directly supported and influenced by Mao's theoretical principles, sought to adopt this strategy of mobilising 'the people'.[11] To do this, they drew upon nationalist and communist ideological sentiments with the goal of recreating Mao's symbiotic relationship between insurgent and populace at a national level. Consequently, counterinsurgents sought to reverse-engineer these principles. To win, McCuen theorised, 'they must not only defeat the revolutionary attempts to mobilize the people, but to mobilize the people

---

8   Mao, *On Guerrilla Warfare*, 47.
9   A. Smedley, 'The Red Phalanx' in Gerard Chaliand (ed.), *Guerrilla Strategies: An Historical Anthology from the Long March to Afghanistan* (London: University of California Press, 1982), 59; Mao, *On Guerrilla Warfare*, 92; Beckett, *Modern Insurgencies and Counter-Insurgencies*, 74.
10   Ernesto 'Che' Guevara, *Guerrilla Warfare* (London: Ocean Press, 2006), 32. This overlooks the centrality of coercion as a tool to both enact and consolidate radical political changes during and after the Civil War. See Frank Dikötter, *The Tragedy of Liberation: A History of the Chinese Revolution, 1945–57* (London: Bloomsbury, 2013), 83. Political scientists have nonetheless studied the role of coercion and violence as a tool of interaction with the population. See Stathis N. Kalyvas, *The Logic of Violence in Civil War* (London: Cambridge University Press, 2006). The tools employed by insurgents to govern and how they deploy an array of techniques is an emergent field of academic study. See Ana Arjona, Nelson Kasfir, and Zachariah Cherian Mampilly (eds), *Rebel Governance in Civil War* (Cambridge: Cambridge University Press, 2015); Nelson Kasfir and Zachariah Cherian Mampilly, 'Is ISIS Good at Governing?', *Brookings* 2016, available at https://www.brookings.edu/blog/markaz/2016/03/22/experts-weigh-in-part-6-is-isis-good-at-governing/, accessed 16 October 2018.
11   Robert Thompson, *Defeating Communist Insurgency: Experiences from Malaya and Vietnam* (London: Chatto & Windus, 1966); Robert Taber, *War of the Flea: A Study of Guerrilla Warfare Theory and Practise* (Washington DC: Brassey's, 2002).

themselves'.[12] The importance of 'the people' thus emerged as a major lesson from engagements in Malaya, Kenya, Algeria and Vietnam, becoming the focal point around which 'classical' COIN doctrine was organised.

The Western-led COIN campaigns in Afghanistan and Iraq during the GWOT revitalised and largely reinforced the conclusions of this body of literature, after the trajectory of both campaigns forced a re-engagement with classical COIN literature. Although the enemies were different and no longer outwardly adhered to communist ideology, the recognition of the importance of demonstrating to the 'population base' that cooperation with the counterinsurgent can best serve their interests remained the central element of COIN theory.[13] The US Army's COIN field manual of 2006 – FM 3-24 – codified these assumptions, conceptualising the insurgency environment as a battleground for the support of 'the people',[14] drawing primarily upon assumptions from the classical COIN literature.

Issues with 'The People'

This overarching conceptualisation of 'the people' is, however, problematic. Firstly, treating 'the people' as a broadly coherent entity oversimplifies societies and the complex relations, cleavages, conflicts, coalitions and actors that make up political order. While the 2006 *FM 3-24* refers to the multi-layered and complex nature of insurgency-affected polities, the task of mapping 'cultural specificities' and 'social structure' remained, as Kalyvas suggests, 'outsourced' to local commanders, rather than engaged with on a paradigmatic or theoretical level.[15] The 2014 edition of *FM 3-24*, addressing the multi-layered nature of popular legitimacy, acknowledges that populations are not monolithic and that insurgencies can obtain support from certain sections of the population and not others, but does not develop upon this, retaining 'popular legitimacy' as the overall goal

---

12 John J. McCuen, *The Art of Counter-Revolutionary War: The Strategy of Counter-Insurgency* (London: Stackpole Books, 1966), 56; Robert Thompson, *Defeating Communist Insurgency*, 57–70; Julian Paget, *Counter-Insurgency Campaigning* (London: Faber & Faber, 1967), 36; Roger Trinquier, *Modern Warfare: A French View of Counterinsurgency* (Westport: Praeger Security International, 2006), 7; Frank Kitson, *Low Intensity Operations: Subversion, Insurgency, Peace-Keeping* (London: Faber & Faber, 1971), 29.

13 David Kilcullen, *Counterinsurgency* (Oxford: Oxford University Press, 2010), 4–10.

14 US Army, *FM 3-24 Counterinsurgency* (University of Chicago Press, 2006), 1:3, available at http://usacac.army.mil/cac2/Repository/Materials/COIN-FM3-24.pdf, accessed 16 October 2018.

15 Stathis N. Kalyvas, 'Review of the New U.S. Army/Marine Corps Counterinsurgency Field Manual', *Perspectives on Politics* 6:2 (2008), 352–53, available at https://doi.org/10.1017/S1537592708081176 accessed 16 October 2018.

for counterinsurgents.[16] According to Smith, 'the people' may not be a monolithic block, but are nonetheless an entity that can be united by political leadership.[17]

The emphasis therefore remains weighted towards the 'the people' rather than the formal or informal actors, institutions and groups within the population around which individual or collective power is located. This is problematic as it risks analytically underestimating the agency of key non-state actors, groupings and coalitions below the level of 'the people'. As conflicts such as those in Iraq and Afghanistan demonstrate, different groupings may occupy different positions of power, influence or closeness to the state, engendering and structuring competitive or conflictual relationships between them. The literature on 'new wars,' ethno-religious and sectarian insurgencies and warlord politics highlights that insurgents and indeed states have, in particular cases, consolidated power based on appeals to specific communities, 'bidding' for power within these communities and then using this power to influence the wider political environment.[18] The Mahdi Army, for example, governed the Shi'a dominated district of 'Sadr City' in Baghdad based on religious appeals, providing protection from Sunni militias, welfare and medical assistance to the community and using this position of influence to assert itself against both rival Shi'a contenders and externally against American forces.[19] Nor are these dynamics restricted to insurgents; following the US withdrawal from Iraq, the political dominance of Shi'a elites and the enactment of exclusionary policies within the central government apparatus served to further alienate Sunni communities, a process which in turn created conducive conditions for IS to exploit and appeal to disaffected Sunni communities in western Iraq.[20] While it must be recognised that these communities are not static entities and can evolve over time, current conceptualisations of the

---

16 US Army, *FM 3-24 Insurgencies and Countering Insurgencies* (Chicago: University of Chicago Press, 2014), 1:10, 1:19, available at http://fas.org/irp/doddir/army/fm3-24. pdf, accessed 16 October 2018.

17 Smith, *The Utility of Force*, 279–80.

18 Mary Kaldor, for example, draws attention to the mobilising potential of identity politics. See Mary Kaldor, *New and Old Wars: Organised Violence in a Global Era* (New Jersey: John Wiley & Sons, 2013). For 'outbidding' within communities, see Andrew H. Kydd and Barbara F. Walter, 'The Strategies of Terrorism', *International Security* 31:1 (2006), 16 and Smitana Saikia, 'General Elections 2014: Ethnic Outbidding and Politics of "Homelands" in Assam's Bodoland', *Contemporary South Asia* 23:2 (2015), 1–12, available at https://doi. org/10.1080/09584935.2015.1029435 accessed 16 October 2018.

19 Andrew Hubbard, 'Plague and Paradox: Militias in Iraq', *Small Wars and Insurgencies* 18:3 (2007), 346–49.

20 Lina Khatib, 'The Islamic State's Strategy: Lasting and Expanding', *Carnegie Middle East Center Blog*, 29 June 2015, available at http://carnegie-mec.org/2015/06/29/islamic-state-s-strategy-lasting-and-expanding-pub-60511, accessed 16 October 2018.

population assume that 'the people' are fundamentally malleable and able to be shaped by the right COIN or insurgent approach.[21] This overlooks the salience and durability of existing forms of politics that may produce different COIN outcomes. Conceptually, population interactions in COIN must therefore take into account the political agency of actors below the amorphous mass of 'the people' from the outset.

Furthermore, the notion of 'the people' analytically flattens out the hierarchical structures and relations within societies. As Kalyvas suggests, in many cases both counterinsurgents and insurgents negotiate with leadership structures within communities that themselves command existing forms of power, influence and legitimacy, rather than directly engaging with populations themselves. The Al-Anbar Awakening in 2007 Iraq, for instance, marked a series of negotiation processes with key community leaders,[22] while hierarchical village headmen and tribal organisations in the Naga Hills of Northeast India, rather than the everyday villager, play a crucial role in forming trust-based relationships between insurgents and counterinsurgents.[23] Counterinsurgents in Afghanistan were compelled to deal with often powerful regional and local warlords that precluded the implementation of a formal statebuilding approach to COIN associated with connecting the state to 'the people',[24] a dilemma that is reflective of the complexities of statebuilding approaches more broadly.

This raises questions concerning COIN theory more broadly and the relationship between winning 'the people' and benchmarks for success and failure in COIN. While widely considered the ultimate objective for counterinsurgents,[25] it is far from clear that counterinsurgents necessarily possess the political will or capacity to implement radical and potentially costly statebuilding campaigns to integrate 'the people' under one political unit. Clearly, the amount of capital that political leaders are willing to spend

---

21 Kalyvas, 'Review of the New U.S. Army / Marine Corps Counterinsurgency Field Manual', 352.

22 Kalyvas, 'Review of the New U.S. Army / Marine Corps Counterinsurgency Field Manual', 352.

23 Namrata Goswami, *Indian National Security and Counter-Insurgency: The Use of Force vs Non-Violent Response* (Abingdon: Routledge, 2015).

24 Dipali Mukhopadhyay, 'Warlords as Bureaucrats: The Afghan Experience' (Washington DC: Carnegie Endowment, 2009), 9–20, available at http://carnegieendowment. org/files/warlords_as_bureaucrats.pdf, accessed 16 October 2018; Andrea M. Lopez, 'Engaging or Withdrawing, Winning or Losing? The Contradictions of Counterinsurgency Policy in Afghanistan and Iraq' in Mark T. Berger and Douglas A. Borer (eds), *The Long War – Insurgency, Counterinsurgency and Collapsing States* (Abingdon: Routledge, 2013), 50; Romain Meljacq, 'Warlords, Intervention, and State Consolidation: A Typology of Political Orders in Weak and Failed States', *Security Studies* 25:1 (2016), 85–110.

25 K. Sepp, 'Best Practices in Counterinsurgency', *Military Review* (2005), 9.

on such a project is finite and contingent upon political considerations.[26] With this in mind, the disaggregation of 'peoples' at a theoretical level provides a framework for examining how counterinsurgents engage in the trade-offs, bargains and negotiations that take place amidst this backdrop of finite political will and resources.

Reconceptualising 'war amongst the people' towards a model of 'war amongst peoples' thus provides an opportunity to elevate this process of socio-political disaggregation at the theoretical level. Conceptualising 'the people' as 'peoples' or entities rather than part of one entity, while recognising that these are by no means fixed, encourages analysts to prioritise mapping COIN approaches onto the fragmented and multi-layered nature of insurgency-affected societies. This should not be confused with the 'primordialist' school of thought within the civil wars literature, which over-ascribes 'hardness' to the boundaries of ethnic and religious communities and engenders a sense of intractability between them. In 'war amongst peoples', who these 'peoples' are does not necessarily refer to ethnic or religious communities, but can instead be defined as the groupings and formations significant to the political life of a state. These may include prominent ethnic or religious communities, but, as the case of Shi'a militias in Iraq demonstrated, these communities may again be fragmented according to political allegiance to a particular militia. Economic and social class backgrounds in cities and allegiance to clan, tribal and village units in rural areas may also play a significant role in hierarchically ordering the relations between 'peoples' in a society. 'War amongst peoples', in calling for the development of robust theoretical linkages between the political complexity of societies and COIN theory, therefore offers an opportunity to theorise how counterinsurgents have sought to negotiate the horizontal and vertical demarcations within societies. In doing so, this allows us to better connect theory to the politics of insurgency and COIN, exploring the processes of contestation, coercion, accommodation and bargaining that take place between counterinsurgents and these different socio-political groupings.

The Syrian Civil War, 2011–17

The Syrian state's approach to the civil war it faced from 2011 represents a particularly intense and complex case in point that serves to draw

---

26 Paul Staniland, 'Counterinsurgency and Violence Management' in Celeste Gventer, David Jones and M.L.R Smith (eds), *The New Counter-Insurgency Era in Critical Perspective* (Basingstoke: Palgrave Macmillan, 2014), 144–56.

out some of these dynamics. COIN campaigns waged by authoritarian regimes such as Russia and Syria are frequently misdiagnosed as 'force-centric' due to the visible, brutal and deliberate employment of standoff firepower in densely-populated areas such as Grozny or Aleppo.[27] According to this line of thinking, such military-centric campaigns do not account for the predominantly political nature of insurgency and COIN, since the counterinsurgent does not employ strategies conducive to winning over 'the people' in their entirety. However, 'authoritarian counterinsurgency comprises far more than overwhelming force', producing state–population interactions in accordance with logics unique to the political considerations of these regimes.[28] Indeed, the Syrian state, while dominated by the Alawite sect at the highest echelons of power, has undertaken a multifaceted process of co-optation, coalition-building, coercion and societal transformation to negotiate the various 'peoples' within the country that ensured a modicum of political stability between 1970–2011.[29] Following the onset of the civil war, the collapse of several of the building blocks underpinning regime power has compelled a recalibration of regime priorities towards the consolidation of alliances with core groupings, including attempts to placate potential allies and appeals to segments of communities from which insurgent groups draw much of their support.

Background: Pre-war State-Society Relations

Syria's estimated 22.5 million population is divided according to both religious and ethno-linguistic affiliations. Sunni Muslims constitute a broad majority (roughly 75 per cent), while prominent religious minorities include the Alawi sect of Shi'a Islam (8–15 per cent), Christianity (10 per cent), Druze (3 per cent) and Ismaili (0.8 per cent).[30] There are close, overlapping intersections between religious, ethno-linguistic, regional and political identity in Syria that have been influenced by

---

27 John Russell, *Chechnya: Russia's War on Terror* (London: Routledge, 2007), xii–11; Witold Mucha, 'Does Counterinsurgency Fuel Civil War? Peru and Syria Compared', *Critical Studies on Terrorism* 6:1 (2013), 140–66.

28 David H. Ucko, '"The People Are Revolting": An Anatomy of Authoritarian Counterinsurgency', *Journal of Strategic Studies* 39:1 (2016), 30, available at https://doi.org/10.1080/01402390.2015.1094390, accessed 16 October 2018.

29 See Raymond Hinnebusch, *Syria: Revolution From Above* (London: Routledge, 2001).

30 'Guide: Syria's Diverse Minorities', *BBC News*, 9 December 2011, available at http://www.bbc.co.uk/news/world-middle-east-16108755, accessed 16 October 2018; Faisal Irshaid, 'Syria's Druze under Threat as Conflict Spreads', *BBC News*, 19 June 2015, available at http://www.bbc.co.uk/news/world-middle-east-33166043, accessed 16 October 2018.

important historical developments. Under Ottoman rule, Syrian society was compartmentalised across different tribal, linguistic and religious groups. The interference of northern imperial powers into the affairs of the Ottoman Empire, in acting as protectors of specific minority communities, drew the ire of the Empire and the Sunni Muslim majority population. After the First World War, the French colonial authorities intervened into and played off tribal leaders, while drawing extensively from the minority Alawi, Druze, Kurdish and Circassian communities for military service, setting the precedent for sectarian competition and over representation within state institutions, such as the armed forces. Following independence in 1946, ideologies such as Arab nationalism and socialism have intersected with and mobilised sectarian loyalties as poor, rural minorities perceived class conflict as a struggle against groups, such as wealthy urban Sunnis.[31] Such tensions were exploited after the Ba'ath Party took power from 1963, especially by its radical wing from 1965, whose reforms attacked 'peoples' that constituted potential opponents while building up a series of new social classes closely associated to the state. These laid the foundations for the president Hafiz al-Asad to later consolidate authority and forge alliances, extending the state's reach into society.[32]

Political Order under Hafiz al-Asad

Syria's political system remained decentralised and consisted of fluctuating and competing elite factions permeating the state's main power institutions: the ruling party, the presidency, the military and the security agencies.[33] Hafiz al-Asad's military coup in 1970 established the dominance of one faction, which was gradually cemented by a combination of coalition-building techniques and the creation of a cult of personality centred around the president to produce everyday compliance.[34] Although the Ba'ath Party and the Syrian state is often associated with the domination of Alawi elites, Hafiz al-Asad co-opted nonthreatening rival parties, facilitated the creation of overlapping interests between regime members and elements of the

---

31 Nikolaos van Dam, *The Struggle for Power in Syria: Politics and Society under Asad and the Ba'th Party* (London: I.B. Tauris, 2011), 2–13; Hinnebusch, *Syria: Revolution From Above*, 3–4.

32 Hinnebusch, *Syria: Revolution From Above*, 47–58; Pete Moore and Bassel Salloukh, 'Struggles under Authoritarianism: Regimes, States, and Professional Associations in the Arab World', *International Journal of Middle East Studies* 39:1 (2007), 66.

33 Joshua Stacher, *Adaptable Autocrats: Regime Power in Egypt and Syria* (Stanford, CA: Stanford University Press, 2012), 13–15.

34 See Lisa Wedeen, *Ambiguities of Domination: Politics, Rhetoric and Symbols of Contemporary Syria* (Chicago: University of Chicago Press, 2015).

bourgeoisie,[35] forged patrimonial alliances with segments of populations across the sectarian, ethnic and religious divides and expanded both the party, the state's coercive apparatus and their level of penetration into society.[36] For Stacher, this gave segments of different 'peoples',

> a stake in the continuity of the regime. He took the mosaic parts of society and bound them together before expanding the regime outwardly into society, particularly into the rural areas. Asad's ability to make deals allowed the political system to rest on key pillars such as the military, ruling party, and security services.[37]

One of the most significant developments in this process was the deliberate co-optation of Sunnis by the regime as part of this broader strategy to forge an elite coalition across sectarian boundaries to nullify the emergence of a coherent front against it.[38] At the core of this was Asad's co-optation of the Sunni business classes, using limited economic liberalisation that 'forged networks of capital that bind elite business actors to state officials'.[39]

However, the 1977–82 Islamist uprising revealed both the vulnerability and the durability of this system. The revolt revealed the potential threat that an opposition movement could pose by mobilising disenfranchised sections of society along sectarian lines, prompting the employment of heavy force against militants in Hama in 1982 that bear striking similarities to the events of 2011.[40] On the other hand, the regime used its political loyalties to prevent the formation of a coherent opposition. The Sunni population remained split during the rebellion and crucial 'peoples', such as the predominantly Sunni business community in Damascus, remained allied to the government following limited liberalisation processes that had benefitted the community.[41] Furthermore, it mobilised its extensive patronage networks to suppress the uprising, raising party-associated

---

35   Adrien Desbonnet, 'Tactical Evolutions in Syria' (Paris: Centre de Doctrine d'Emploi des Forces, 2015), 13, available at https://www.cdec.terre.defense.gouv.fr/contents-in-english/our-publications/cahier-du-retex/cahier-du-retex-research/tactical-evolutions-in-syria-2011-2014, accessed 16 October 2018.

36   Hinnebusch argues that it is problematic to conceptualise the Syrian regime as 'exclusively an Alawi one or that Alawi dominance translated exclusively into a politics of sectarian privilege and rivalry'. Hinnebusch, *Syria: Revolution From Above*, 65–91, 110.

37   Stacher, *Adaptable Autocrats*, 13.

38   Hinnebusch, *Syria: Revolution From Above*, 70.

39   Bassam Haddad, 'The Syrian Regime's Business Backbone', *Middle Eastern Report* 42: 262 (2012), available at http://www.merip.org/mer/mer262/syrian-regimes-business-backbone, accessed 16 October 2018.

40   Dara Conduit, 'The Patterns of Syrian Uprising: Comparing Hama in 1980–1982 and Homs in 2011', *British Journal of Middle Eastern Studies* 44:1 (2017), 73–87.

41   Hinnebusch, *Syria: Revolution From Above*, 98; Anthony Shadid, 'Rami Makhlouf becomes magnet for Syrian Dissent', *The New York Times*, 30 April 2011, available at https://www.

militias and deploying elite armed units under the command of trusted political allies and family members.[42] Upon Hafiz al-Asad's death in 2000, his successor, Bashar al-Asad, inherited a well-oiled regime apparatus welded together by elaborate systems of patronage and coercion.

Bashar al-Asad and the Onset of the Civil War

The outbreak of the Syrian Civil War in 2011, although triggered by the exogenous dynamics brought about by the Arab Spring, came as a result of the long-term erosion and disruption of the delicate balance between the resources available to the state and its management of 'peoples'.[43] The end of Cold War largesse, economic stagnation during the 1990s and depleting oil reserves prompted an acceleration of economic liberalisation.[44] This process primarily advanced the interests of business elites associated with the regime,[45] but also benefitted new and traditional urban Sunni economic communities in cities such as Aleppo and Damascus, giving them little incentive to vigorously support the uprising.[46] This process of 'embourgeoisiement' nonetheless created 'losers'.[47] In jettisoning its 'social market' principles, economic liberalisation generated huge levels of unemployment and alienated vast sections of the party's rural peasant base and poorer urban communities, creating a conducive base for protest and armed rebellion.[48] This demonstrates the fragility of attempts to manage 'peoples' and the ease with which the erosion of such equilibria can create the conditions for armed conflict, creating considerable hurdles for counterinsurgents that need to engage with this process while, at the same time, engaging with parallel attempts to do so from rival rebel and external actors.

nytimes.com/2011/05/01/world/asia/01makhlouf.html, accessed 20 June 2018; Thomas Pierret, 'The Syrian Baath Party and Sunni Islam: Conflicts and Connivance', *Crown Center for Middle East Studies Middle East Brief* (2014), 3, available at https://www.brandeis.edu/crown/publications/meb/MEB77.pdf, accessed 16 October 2018.

42   Desbonnet, 'Tactical Evolutions in Syria', 11, 26; Eva Bellin, 'The Robustness of Authoritarianism in the Middle East: Exceptionalism in Comparative Perspective', *Comparative Politics* 36:2 (2004), 149.

43   Hinnebusch, *Syria: Revolution From Above*, 103; Volker Perthes, *Syria under Bashar Al-Asad: Modernisation and the Limits of Change*, Adelphi Paper 366, 8 (London: International Institute for Strategic Studies, 2004).

44   Perthes, 'Syria under Bashar Al-Asad', 29–33; Hinnebusch, *Syria: Revolution From Above*, 103; Hussein Almohamad and Andreas Dittmann, 'Oil in Syria between Terrorism and Dictatorship', *Social Sciences* 5:2 (2016), 5.

45   See for example Shadid, 'Rami Makhlouf becomes magnet for Syrian Dissent'.

46   Pierret, 'The Syrian Baath Party and Sunni Islam: Conflicts and Connivance', 5.

47   Hinnebusch, *Syria: Revolution From Above*, 9.

48   Desbonnet, 'Tactical Evolutions in Syria', 4; Hinnebusch, 'Syria', 796.

'Managing' the 'War Amongst Peoples' in Syria

The intensification of the civil war has forced the Syrian regime to recalibrate its priorities away from restoring its dominant position atop society towards survival. The regime lost vast northern and eastern portions of its territory and lost 20–30 per cent of its armed forces through defections.[49] While the revolt signalled the state's failure to maintain the delicate equilibrium of bargains with 'peoples' across Syrian society, the regime, supported to a considerable extent by its Iranian and Russian allies, has nonetheless held together enough of the 'peoples' underpinning its core alliance base to ensure a form of survival. It has done this through employing sectarian rhetoric to instil fear of what is portrayed as a jihadist opposition, using massive force against rebel-held population centres to isolate insurgents from the population and deter further defections, while forging tenuous tactical relationships with those beyond its core base to ensure survival.

Sectarian Rhetoric

The 'peoples' upon which the regime has historically depended, such as its core Alawi community and the Christian, Druze and Ismaili minorities, are by no means unified communities, nor are they unanimous or indeed set in their support for Asad. However, the regime has been able to shore up its patrimonial deal-making with sectarian rhetoric to secure these communities from large-scale defections. From the outset of the conflict, the Asad regime has used fear as an effective weapon, using its well-oiled propaganda machinery to exploit the fears of Sunni militant reprisals that are entrenched in the political mindsets of minority communities.[50] This has had the effect of cementing the bonds between these communities and the regime, by fear of the possible alternatives if not by affinity. In Alawi-dominated regions, for example, there is a reluctant recognition that 'fighting to protect Assad has become synonymous with fighting for political, financial, and social power for potentially decades to come'.[51] The Asad regime has, for the most part, been able to rely on communities such as Alawites and Christians to

---

49   Desbonnet, 'Tactical Evolutions in Syria', 228.

50   Desbonnet, 'Tactical Evolutions in Syria', 44; Joseph Holliday, *The Assad Regime: From Counterinsurgency to Civil War*, (Washington DC: Institute for the Study of War, 2013), 18, available at http://www.worldaffairsjournal.org/content/assad-regime-counterinsurgency-civil-war, accessed 16 October 2018.

51   Jomana Qaddour, 'Unlocking the Alawite Conundrum in Syria', *The Washington Quarterly* 36:4 (2013), 69–70; Lauren Williams, 'Syria's Alawites Not Deserting Assad yet, despite Crackdown', *Middle East Eye*, 11 September 2014, available at http://www.

raise 'popular committee' militias in their areas of jurisdiction, performing defensive duties and cooperating with regime forces,[52] but has had to carefully manage this support. Tellingly, Damascus was compelled to agree to the exclusive deployment of Druze conscripts in their home areas after aggressive recruitment efforts provoked violent protests in 2015, even as Al Qaeda affiliate Jabhat al-Nusra advanced on Druze-inhabited territories in south-western Syria.[53] Clearly, while sectarian fears have held the minority coalition together, an alteration of the power calculus has taken place as these communities become increasingly central to regime survival. At the same time, involvement of allied proxies, such as Iran and Hezbollah, has allowed these actors to consolidate their own power bases within Syria that, while convergent with Syrian interests, are intended to fulfil their own regional interests rather than those of Damascus, again demonstrating the trade-offs that the regime has been required to make over control of the Syrian population.[54]

Cross-Sectarian Bargaining

Parallel to the government's own sectarian rhetoric, however, lies the spectre of the collapse of one of the central pillars of the cross-sectarian coalition established under Hafiz and Bashar al-Asad – the relationship between the state and the urban Sunni business classes. With the collapse of Syria's wider economy and the onset of sanctions, they have become increasingly crucial to regime survival and have to a great extent remained wedded to the regime's interests. The intensification of the conflict has undoubtedly seen businesses relocate to neighbouring countries; the regime's lawsuit submitted to a French court in 2015 accusing Turkey of 'stealing' factories is reflective of the significance it ascribes to the business classes.[55] It has also had to tolerate increased levels of non-cooperation or protest as it grows increasingly dependent on the business classes. In 2012, for example, business communities in Damascus conducted

middleeasteye.net/in-depth/features/syrias-alawites-not-deserting-assad-yet-despite-crackdown-526622504, accessed 16 October 2018; Leon Goldsmith, *Cycle of Fear: Syria's Alawites in War and Peace* (Oxford: Oxford University Press, 2015).

52 Holliday, 'The Assad Regime: From Counterinsurgency to Civil War', 18; Desbonnet, 'Tactical Evolutions in Syria', 27.

53 Irshaid, 'Syria's Druze under Threat as Conflict Spreads'.

54 Will Fulton, Joseph Holliday and Sam Wyer, 'Iranian Strategy in Syria' (Washington DC: Institute for the Study of War, 2013), available at http://www.understandingwar.org/sites/default/files/IranianStrategyinSyria-1MAY.pdf, accessed 16 October 2018.

55 Younes Ahmed, 'Syrian Factories Relocating to Coastal Area', *The Arab Weekly*, 26 June 2015, available at http://www.thearabweekly.com/Opinion/887/Syrian-factories-relocating-to-coastal-area, accessed 16 November 2018.

shutdowns protesting the massacre of 108 civilians by pro-state militias in Houla during May 2012.[56] The regime has consistently presented itself as an ally to those who desire stability, juxtaposing this with its propaganda labelling rebel groups as jihadists and using this as a tool to maintain compliance among businesses and the Sunni middle classes.[57] This has helped secure the neutrality of many businesses, including those that have relocated abroad, and defections have remained minimal.[58] Even as Damascus shifted towards a scorched earth strategy against former business hubs such as Aleppo, it actively supported the relocation of businesses away from such cities, moving 109 factories during 2013–14 to coastal areas, such as Latakia and Tartus. This offered businesses a modicum of stability given these areas' status as Alawi-dominated strongholds,[59] with the added effect of further binding these communities to the regime's interests.

Furthermore, Sunnis remain a large percentage of the armed forces; Syrian Air Force recruitment as of 2013 remained rooted in Sunni upper and middle-class communities.[60] Most of the army's 15–20 per cent defections were from the lower-ranking ranks of Sunni infantrymen. Although such defections have slowed since 2012, only roughly 65,000 out of a total of 220,000 armed forces personnel were, as of 2015, considered trustworthy by the regime.[61] The Syrian government has relied on selective deployments led by key Alawite allies for the prosecution of offensive military operations. The control of key armoured and special forces units has provided the regime with a reliable offensive force, while the under-utilisation of potentially less-reliable Sunni-dominated units has hedged against the risk of defections.[62] The state has also drawn upon criminalised

56 Khaled Yacoub Oweis, 'Damascus Merchants Strike over Houla Killings', *Reuters*, 28 May 2012, available at http://www.reuters.com/article/uk-syria-unrest-strike-idUKBRE84R0MC20120528, accessed 16 October 2018.

57 Desbonnet, 'Tactical Evolutions in Syria', 58.

58 Aurora Sottimano, 'The Syrian Business Elite: Patronage Networks and War Economy', *Syria Untold*, 24 September 2016, available at http://www.syriauntold.com/en/2016/09/the-syrian-business-elite-patronage-networks-and-war-economy/, accessed 16 October 2018.

59 Ahmed, 'Syrian Factories Relocating to Coastal Area'.

60 Reva Bhalla, 'Making Sense of the Syrian Crisis', *Stratfor Worldview*, 5 May 2011, available at https://www.stratfor.com/weekly/20110504-making-sense-syrian-crisis, accessed 16 October 2018; Aymenn Jawad, 'Is an Alawite State Assad's Plan B?', *The Independent*, 7 January 2013, available at http://www.independent.co.uk/voices/comment/is-an-alawite-state-assads-plan-b-8441059.html, accessed 16 October 2018.

61 Desbonnet, 'Tactical Evolutions in Syria', 228.

62 Brian Michael Jenkins, *The Dynamics of Syria's Civil War* (Santa Monica: RAND Corporation, 2014), 13, available at http://www.rand.org/pubs/perspectives/PE115.html, accessed 16 October 2018; Holliday, 'The Assad Regime: From Counterinsurgency to Civil War', 11–15.

shabiha militias operated by both Asad allies with family ties, as well as non-Alawite criminal gangs. These are not under the strict control of the state and offer the potential for large-scale enrichment in return for political loyalty to Asad.[63] The government also received a significant political boost following Russian intervention into the conflict in September 2015, which has provided substantial military reinforcement to boost the regime's capacities for conducting its broader COIN strategy.

'Peoples' in Rebel-Held Areas

The direction of regime COIN operations themselves has been inextricably tied to the political priorities of regime survival; the targeting of major population centres, while allowing vast parts of rural Syria to go unchallenged, demonstrates that the regime has been determined to secure the key sites in which it has fused political and economic power amongst its elite coalition. Though its tactics have been extremely brutal in nature, the approach appears to serve two fundamentally political functions that are essentially predicated upon fear. Firstly, it has driven wedges between rebels and 'peoples' in rebel-held territory, while it also communicates power and control to the core of its own coalition of 'peoples'. During the early phases of the war, Syrian COIN operations initially mirrored the campaign waged against Islamist militants during 1977–82, employing standoff weaponry to bombard insurgent-held zones and then using politically-loyal militias to clear areas. However, as state capacities thinned, discouraging the use of infantry to 'clear' areas,[64] this strategy appeared to shift during the battle of Aleppo (2012–16) towards one of 'wear[ing] out the population, preventing it from joining the insurgency by capitalizing on fear and tying their survival to fighting coming to an end'.[65] This strategy has focused on the prolonged use of standoff weaponry, such as artillery and aircraft in cities such as Aleppo, deliberately targeting the infrastructure that sustains civilian life, such as hospitals and bakeries, and eroding the political, psychological and physical will of people living in rebel-held areas.[66] Clearly, the state

63 Holliday, 'The Assad Regime: From Counterinsurgency to Civil War', 17–18; Harriet Alexander, 'The Shabiha: Inside Assad's Death Squads', *The Daily Telegraph*, 2 June 2012, available at http://www.telegraph.co.uk/news/worldnews/middleeast/syria/9307411/The-Shabiha-Inside-Assads-death-squads.html, accessed 16 October 2018.
64 Holliday, 'The Assad Regime: From Counterinsurgency to Civil War', 20.
65 Desbonnet, 'Tactical Evolutions in Syria', 41–45.
66 Annia Ciezadlo, 'The War on Bread', *The New Republic*, 15 February 2014, available at https://newrepublic.com/article/116615/syrian-war-crimes-regime-bombs-bakeries-uses-starvation-weapon, accessed 16 October 2018. Estimates from aid agencies suggest that at least 80 per cent of the country's hospitals were attacked during 2016. See Bel Trew,

does not intend to 'win' populations in rebel-held territory during these sustained and deliberate siege operations. The strategy deliberately aims to divide populations in rebel-held areas, linking rebel presence to large-scale suffering and, in doing so, disrupting rebel efforts to hold onto constituents through means such as service provision and split the regime from its constituents. Rebels' inability to provide services under such circumstances of pressure served to turn civilians against them in newly-captured rebel areas in Aleppo.[67]

This strategy of making life physically intolerable in rebel-held areas reinforces the regime's central propaganda message: that support for rebels is synonymous with chaos and destruction, while compliance with the regime offers order, stability and physical survival. This point was illustrated when a series of ceasefires signed in and around Damascus in 2013–14 restored public services to those affected by hunger. Furthermore, the restoration of state provisions, in return for symbolic gestures, such as the raising of the Syrian flag, secured a significant propaganda victory for the state.[68] Such propaganda, which has also included visual displays of normality and vibrancy in government-held Aleppo,[69] communicates state control and power to 'peoples' already under its control. Such forms of propaganda mark al-Asad's direct continuation of the techniques employed by his father's cult of personality, as argued by Wedeen that, while filled with often perplexing claims, nonetheless reproduced and sustained political domination without pretensions to securing popular legitimacy.[70]

'Assad's Forces "Deliberately Attacking Hospitals"', *The Times*, 3 April 2017, available at https://www.thetimes.co.uk/article/assads-forces-deliberately-attacking-hospitals-jr69wfq73, accessed 16 October 2018.

67 Indeed, Holliday suggests that the regime's use of 'barrel bombs' and militia-led massacres can be situated within the context of a 'depopulation' strategy. See Holliday, 'The Assad Regime: From Counterinsurgency to Civil War', 23–24.

68 Desbonnet, 'Tactical Evolutions in Syria', 44–45; 'Syrian Rebel-Held Town 'Raises Government Flag in Exchange for Food', *The Daily Telegraph*, 27 December 2013, available at http://www.telegraph.co.uk/news/worldnews/middleeast/syria/10538769/Syrian-rebel-held-town-raises-government-flag-in-exchange-for-food.html, accessed 16 October 2018.

69 In one particularly bizarre piece by Syria's Tourism Ministry, a drone flies over government-occupied parts of Aleppo amid a backdrop of music to the tune of the opening credits for HBO series 'Game of Thrones'. See Syrian Ministry of Tourism, *Aleppo Will of Life* (Damascus: Syrian Ministry of Tourism, 2016), available at https://www.youtube.com/watch?v=4DWLUhoW5O0, accessed 16 October 2018.

70 Annia Ciezadlo, 'Why Assad's Propaganda Isn't as Crazy as it Seems', *NewsDeeply*, 3 October 2016, available at https://www.newsdeeply.com/syria/articles/2016/10/03/analysis-why-assads-propaganda-isnt-as-crazy-as-it-seems accessed 16 October 2018; Wedeen, *Ambiguities of Domination*, xii, 156.

External Actors and Constituencies

While it cannot utilise sustained narratives to reproduce dominance and control in the same way in an international context, there is evidence that the regime has recognised the importance of international public opinion and has sought to repackage its key narratives to western audiences in an attempt to exploit fears of IS terrorism.[71] Both Bashar al-Asad and his British-born wife, Asma al-Asad, have conducted interviews to international media houses in which they have conveyed messages to western audiences portraying the conflict as a dichotomous one between secularism and terrorism,[72] while the Syrian propaganda narrative has been reinforced and sustained by Russian foreign media networks, such as RT.[73]

Finally, the regime has also sought to reach out to 'peoples' for the purposes of tactical utility and has most notably developed a tacit, if shaky, understanding with the Kurdish Democratic Union Party (PYD). In 2011, the Syrian state granted citizenship to an estimated 150,000–300,000 Kurds in the Hasakah region in an attempt to alleviate growing political pressure.[74] The bulk of regime forces then peacefully withdrew from Kurdish areas in June 2012, leaving a minor administrative and security presence to retain small pockets of influence. These forces largely avoided clashes with the PYD, while Syrian and Kurdish forces have even cooperated at key stages,[75]

71 Kareem Chehayeb, 'Syrian State Media: War Propaganda, Whitewashing and Denial', *Al Araby*, 12 October 2016, available at https://www.alaraby.co.uk/english/comment/2016/10/13/syrian-state-media-war-propaganda-whitewashing-and-denial, accessed 16 October 2018.

72 'Syria Conflict: BBC Exclusive Interview with President Bashar Al-Assad', *BBC News*, 9 February 2015, available at https://www.youtube.com/watch?v=yiC4w7Erz8I, accessed 16 October 2018.

73 'It's Clear Assad Has Won This War' – Former US Ambassador', *RT*, 11 December 2016, available at https://www.youtube.com/watch?v=ZzcDf17MQU4, accessed 16 October 2018; 'Syria's First Lady Asma Al-Assad's Interview with Russia's Channel 24', *Channel 24*, 18 October 2016, available at https://www.youtube.com/watch?v=-Du75qd8Pgk, accessed 16 October 2018; 'Assad to RT: 'I'm Not Western Puppet – I Live and Die in Syria', *RT*, 8 November 2012, available at https://www.youtube.com/watch?v=pdH4JKjVRyA, accessed 16 October 2018.

74 'Syria's Assad Grants Nationality to Hasaka Kurds', *BBC News*, 7 April 2011, available at https://www.bbc.co.uk/news/world-middle-east-12995174, accessed 16 October 2018.

75 International Crisis Group, *Flight of Icarus? The PYD's Precarious Rise in Syria* (Brussels: International Crisis Group, 2014), 7–9 available at https://www.justice.gov/sites/default/files/pages/attachments/2015/05/29/icg_050814.pdf, accessed 16 October 2018; Sam Hamad, 'YPG and Assad: Pragmatic Allies but Unwilling Bedfellows?', *Al Araby*, 26 August 2016, available at https://www.alaraby.co.uk/english/comment/2016/8/26/ypg-and-assad-pragmatic-allies-but-unwilling-bedfellows, accessed 16 October 2018; Tristan Dunning, 'Military Marriages of Convenience Face Uncertain Futures in Syria', *Al Araby*, 5 April 2017, available at https://www.alaraby.co.uk/english/comment/2017/4/10/military-marriages-of-convenience-face-uncertain-futures-in-

such as during the battle to retake Hasakah city in northeast Syria from IS in the summer of 2015.[76] While such alliances have pragmatic, short-term utility, critically, they have evolved primarily in accordance with the battlefield situation. The PYD's consolidation at the expense of the regime in the same city in August 2016 illustrates the unpredictability of such wartime bargaining processes, underscoring the vulnerability of the regime's isolated north-eastern pockets.[77] Similarly, the presence of radical jihadist groups, such as IS, reinforces the regime's central propaganda messages portraying non-regime forces as jihadist terrorists and may also help explain why it reportedly played a significant role in aiding groups such as IS and Jabhat al-Nusra by releasing scores of Islamist prisoners in 2011. There have been reported instances in which regime and IS interests have overlapped, producing examples of tacit restraint and even collusive transactions between regime and IS forces. However, these appear to have been primarily tactical in nature;[78] the 'race to Raqqa' between regime forces and Syrian Democratic Forces (SDF) from 2016–17 highlights a changed battlefield context in which regime advances against IS offer the potential of attaining significant political and symbolic capital to integrate within its broader political messages.[79]

Conclusion

The Syrian case study stands out in its brutality, leading commentators to suggest that Damascus is pursuing a fundamentally 'enemy-centric' COIN strategy.[80] Indeed, if the Syrian Civil War is conceptualised as

syria?utm_campaign=magnet&utm_source=article_page&utm_medium=related_articles, accessed 16 October 2018.

76   Agence France-Presse, 'Syria Army, Kurds Push Islamic State out of Hasakeh City', *NDTV*, 29 July 2015, available at http://www.ndtv.com/world-news/syria-army-kurds-push-islamic-state-out-of-hasakeh-city-report-1201578, accessed 16 October 2018.

77   Rodi Said and Tom Perry, 'Syria Kurds Win Battle with Government, Turkey Mobilizes against Them', *Reuters*, 23 August 2016, available at http://www.reuters.com/article/us-mideast-crisis-syria-kurds-idUSKCN10Y127, accessed 16 October 2018.

78   Michael Becker, 'When Terrorists and Target Governments Cooperate: The Case of Syria', *Perspectives on Terrorism* 9:1 (2015) available at http://www.terrorismanalysts.com/pt/index.php/pot/article/view/404, accessed 16 October 2018.

79   'Syrian Army Advances West of Raqqa: Hezbollah Military Media Unit', *Reuters*, 13 June 2017, available at http://www.reuters.com/article/us-mideast-crisis-syria-army-idUSKBN1941QO, accessed 16 October 2018; Joseph V. Micallef, 'The Race to Raqqa: The Next Russian-American Proxy Battle?', *Huffington Post*, 29 May 2016, available at http://www.huffingtonpost.com/joseph-v-micallef/the-race-to-raqqa-the-nex_b_10193766.html, accessed 16 October 2018.

80   Ben Rich, 'Sticks over Carrots: The Rationale of Assad's Counterinsurgency "Madness"', *The Conversation*, 18 November 2013, available at http://theconversation.com/sticks-

a 'war amongst the people' and, by extension, frames the goals of the counterinsurgent in accordance with securing legitimacy amongst 'the people' – in this case the broad base of the Syrian population – the analysis points towards the regime's dramatic failure in COIN.

Yet, applying these benchmarks to measure success and failure fails to capture how it is that the regime has survived and maintained a support base, despite the collapse of government authority in large portions of the country. As this chapter has argued, Damascus has employed a range of approaches with 'peoples' it considers politically significant, shoring up and maintaining its main elite-based political alliances while attacking and playing off those under the influence of its opponents. It has done this by deploying deliberately sectarian rhetoric, demonising Sunni jihadists to shore up support from increasingly sceptical, but ultimately fearful, minority communities. It has also continued to empower Alawite allies, both within the military and in external militias, turning a blind eye to illicit activities in exchange for loyalty. At the same time, it has bargained to maintain its traditional alliance with Sunni middle classes in a bid to maintain the cross-sectarian alliances crucial to regime survival. In a very different approach, it has also deliberately targeted civilian infrastructure in rebel-held areas in a bid to weaken populations to the extent that they turn against rebels, while the provision of humanitarian aid to these populations following ceasefires serves to reinforce perceptions of regime power in its core constituencies. The Syrian state has essentially engaged with this array of political constituencies based on the historical precedent of Hafiz al-Asad's patronage-based bargaining and deal-making that ensured regime stability from 1970 until the outbreak of the civil war in 2011. However, with the onset of a serious challenge to state authority, the regime's efforts to secure political survival have compelled it to carefully manage these relationships, balancing them alongside tenuous tactical relationships with militant groups emanating from 'peoples' that constitute the active opposition, such as the YPD and, on occasion, IS.

At the time of writing, the effectiveness of the Syrian government's approach to the civil war remains unclear, nor is it certain whether the regime's durability is owed to its own political strategies or the interventions of allies, such as Iran and Russia. However, by examining recent developments in the conflict, the chapter has illustrated the problems existing COIN theory faces when accounting for how the Syrian state has approached engaging with different components of 'the people' during its

---

over-carrots-the-rationale-of-assads-counterinsurgency-madness-20009, accessed 16 October 2018.

COIN campaign. In doing so, it has argued that current conceptualisations of 'war amongst the people' do not adequately capture how states can bargain and negotiate with different communities to fulfil political priorities as tools of COIN, highlighting how COIN theory struggles to capture the constraints and opportunities available to states with differing capacities, motives and interests.[81]

What is needed is a reconceptualisation of 'war amongst the people' towards that of 'war amongst peoples' in an attempt to better connect COIN theory with the evolving politics and dynamism of insurgency and COIN. This should not, however, be interpreted as a call to dismiss the utility of generalisable approaches to studying COIN. Politically disaggregating 'the people' at the outset instead provides us with opportunities to develop novel analytical frameworks to map out the ways in which social forces can stratify, hierarchically structure and order the relations between 'peoples' in a society, from which insights can be drawn from the broader literature on statebuilding, state-society relations and comparative politics. Future research should therefore examine how states engage with the 'war amongst peoples' by conducting comparisons between different 'types' of state, according to capacity and regime type, as well as the different 'types' of COIN in the literature, such as the range of 'enemy-centric' and 'population-centric' approaches. Such comparative research would help to determine whether different 'types' of states and their COIN strategies are or should be more inclined to adopt more divisive or integrative attempts to manage the 'war amongst peoples' to secure their key interests. Other potential avenues to explore within this context could concern how well particular strategies serve particular interests, using 'war amongst peoples' as a tool to shine new light on existing debates concerning the utility of force, inducement and control in COIN.

For practitioners with extensive involvement in COIN, such insights are likely to provide further conceptual underpinnings for already existing efforts to map the social and political terrain of insurgency environments; those across the spectrum of COIN practice are often intimately involved in the processes of balancing, mediating, negotiating and bargaining with the different constituent actors that make up the social and political terrain. Yet these complex processes, often 'outsourced' to local commanders,[82] should be placed at the very centre of thinking and doing COIN, both in military and civilian circles, to encourage greater introspection and

---

81   Staniland, 'Counterinsurgency and Violence Management'.
82   Kalyvas, 'Review of the New U.S. Army / Marine Corps Counterinsurgency Field Manual', 252–53.

learning. Thereafter, a realistic attempt can be made to shape the complex webs of power relations counterinsurgents intervene into, allowing for the development of more finely-tuned indicators gauging 'success' and 'failure' in attempts to manage 'wars amongst 'peoples''.

References

Almohamad, H., and Dittmann, A., 'Oil in Syria between Terrorism and Dictatorship', *Social Sciences* 5:2 (2016).

Arjona, A., Kasfir, N., Mampilly, Z.C. (eds), *Rebel Governance in Civil War* (Cambridge: Cambridge University Press, 2015).

Becker, M., 'When Terrorists and Target Governments Cooperate: The Case of Syria', *Perspectives on Terrorism* 9:1 (2015).

Beckett, I., *Modern Insurgencies and Counter-Insurgencies: Guerrillas and Their Opponents since 1750* (New York: Routledge, 2001).

Bellin, E., 'The Robustness of Authoritarianism in the Middle East: Exceptionalism in Comparative Perspective', *Comparative Politics* 36:2 (2004).

Berger, M. and Borer, D. (eds), *The Long War – Insurgency, Counterinsurgency and Collapsing States* (Abingdon: Routledge, 2013).

Callwell, C.E., *Small Wars: Their Principles and Practice* (London: His Majesty's Stationery Office, 1906).

Chaliand, G., (ed.), *Guerrilla Strategies: An Historical Anthology from the Long March to Afghanistan* (London: University of California Press, 1982).

Chehayeb, K., 'Syrian State Media: War Propaganda, Whitewashing and Denial', *Al Araby*, 12 October 2016, available at https://www.alaraby.co.uk/english/comment/2016/10/13/syrian-state-media-war-propaganda-whitewashing-and-denial, accessed 16 October 2018.

Ciezadlo, A., 'The War on Bread', *The New Republic*, 15 February 2014 available at https://newrepublic.com/article/116615/syrian-war-crimes-regime-bombs-bakeries-uses-starvation-weapon, accessed 16 October 2018.

Ciezadlo, A., 'Why Assad's Propaganda Isn't as Crazy as it seems', *NewsDeeply*, 3 October 2016, available at https://www.newsdeeply.com/syria/articles/2016/10/03/analysis-why-assads-propaganda-isnt-as-crazy-as-it-seems, accessed 16 October 2018.

Conduit, D., 'The Patterns of Syrian Uprising: Comparing Hama in 1980–1982 and Homs in 2011', *British Journal of Middle Eastern Studies* 44:1 (2017).

Desbonnet, A., 'Tactical Evolutions in Syria' (Paris: Centre de Doctrine d'Emploi des Forces, 2015).

Dikötter, F., *The Tragedy of Liberation: A History of the Chinese Revolution, 1945–57* (London: Bloomsbury, 2013).

Fulton, W., Holliday, J. and Wyer, S., 'Iranian Strategy in Syria' (Washington DC: Institute for the Study of War, 2013).

Goldsmith, L., *Cycle of Fear: Syria's Alawites in War and Peace* (Oxford: Oxford University Press, 2015).

Goswami, N., *Indian National Security and Counter-Insurgency: The Use of Force vs Non-Violent Response* (Abingdon: Routledge, 2015).

Guevara, E., *Guerrilla Warfare* (London: Ocean Press, 2006).

Gventer, C., Jones, D., and Smith, M.L.R (eds), *The New Counter-Insurgency Era in Critical Perspective* (Basingstoke: Palgrave Macmillan, 2014).

Gwynn, C.W., *Imperial Policing* (London: Macmillan, 1934).

Haddad, B., 'The Syrian Regime's Business Backbone', *Middle Eastern Report* 42:262 (2012), available at http://www.merip.org/mer/mer262/syrian-regimes-business-backbone, accessed 16 October 2018.

Hinnebusch, R., *Syria: Revolution From Above* (London: Routledge, 2001).

Hinnebusch, R., 'Syria' in Ellen Lust (ed.), *The Middle East* (Thousand Oaks: CQ Press, 2017).

Holliday, J., *The Assad Regime: From Counterinsurgency to Civil War* (Washington DC: Institute for the Study of War, 2013).

Hubbard, A., 'Plague and Paradox: Militias in Iraq', *Small Wars and Insurgencies* 18:3 (2007).

International Crisis Group, *Flight of Icarus? The PYD's Precarious Rise in Syria* (Brussels: International Crisis Group, 2014).

Jenkins, B.M., *The Dynamics of Syria's Civil War* (Santa Monica: RAND Corporation, 2014).

Kaldor, M., *New and Old Wars: Organised Violence in a Global Era* (New Jersey: John Wiley & Sons, 2013).

Kalyvas, S., *The Logic of Violence in Civil War* (London: Cambridge University Press, 2006).

Kalyvas, S., 'Review of the New U.S. Army/Marine Corps Counterinsurgency Field Manual', *Perspectives on Politics* 6:2 (2008).

Kasfir, N., and Mampilly, Z.C., 'Is ISIS Good at Governing?', *Brookings* 2016, available at https://www.brookings.edu/blog/markaz/2016/03/22/experts-weigh-in-part-6-is-isis-good-at-governing/, accessed 16 October 2018.

Khatib, L., 'The Islamic State's Strategy: Lasting and Expanding', *Carnegie Middle East Center Blog*, 2015, available at http://carnegie-mec.org/2015/06/29/islamic-state-s-strategy-lasting-and-expanding-pub-60511, accessed 16 October 2018.

Kilcullen, D., *Counterinsurgency* (Oxford: Oxford University Press, 2010).

Kitson, F., *Low Intensity Operations: Subversion, Insurgency, Peace-Keeping* (London: Faber, 1971).

Kydd, A. and Walter, B., 'The Strategies of Terrorism', *International Security* 31:1 (2006).

Lopez, A.M., 'Engaging or Withdrawing, Winning or Losing? The Contradictions of Counterinsurgency Policy in Afghanistan and Iraq' in Mark T. Berger and Douglas A. Borer (eds), *The Long War – Insurgency, Counterinsurgency and Collapsing States* (Abingdon: Routledge, 2013).

McCuen, J., *The Art of Counter-Revolutionary War: The Strategy of Counter-Insurgency* (London: Stackpole Books, 1966).

McKenzie, A., '"New Wars" Fought "Amongst the People": "Transformed" by Old Realities?', *Defence Studies* 11:4 (December 2011).

Mao Tse-Tung, *On Guerrilla Warfare* (Washington, DC: Presidio Press, 1991).

Meljacq, R., 'Warlords, Intervention, and State Consolidation: A Typology of Political Orders in Weak and Failed States', *Security Studies* 25:1 (2016).

Moore, P., and Salloukh, B., 'Struggles under Authoritarianism: Regimes, States, and Professional Associations in the Arab World', *International Journal of Middle East Studies* 39:1 (2007).

Mucha, W., 'Does Counterinsurgency Fuel Civil War? Peru and Syria Compared, *Critical Studies on Terrorism* 6:1 (2013).

Mukhopadhyay, D. 'Warlords as Bureaucrats: The Afghan Experience' (Washington DC: Carnegie Endowment, 2009), 9–20, available at http://carnegieendowment. org/files/warlords_as_bureaucrats.pdf, accessed 16 October 2018.

Paget, J., *Counter Insurgency Campaigning* (London: Faber & Faber, 1967).

Perthes, V., *Syria under Bashar Al-Asad: Modernisation and the Limits of Change*, Adelphi Paper 366, 8 (London: International Institute for Strategic Studies, 2004).

Pierret, T., 'The Syrian Baath Party and Sunni Islam: Conflicts and Connivance', *Crown Center for Middle East Studies Middle East Brief* (2014) available at https://www.brandeis.edu/crown/publications/meb/MEB77.pdf, accessed 16 October 2018.

Qaddour, J., 'Unlocking the Alawite Conundrum in Syria', *The Washington Quarterly* 36:4 (2013).

Rich, B., 'Sticks over Carrots: The Rationale of Assad's Counterinsurgency "Madness"', *The Conversation*, 18 November 2013, available at http://theconversation.com/ sticks-over-carrots-the-rationale-of-assads-counterinsurgency-madness-20009, accessed 16 October 2018.

Russell, J., *Chechnya: Russia's War on Terror* (London: Routledge, 2007).

Saikia, S., 'General Elections 2014: Ethnic Outbidding and Politics of "Homelands" in Assam's Bodoland', *Contemporary South Asia* 23:2 (2015).

Sepp, K., 'Best Practices in Counterinsurgency', *Military Review* May/June (2005).

Smedley, A. 'The Red Phalanx' in Gerard Chaliand (ed.), *Guerrilla Strategies: An Historical Anthology from the Long March to Afghanistan* (London: University of California Press, 1982).

Smith, R., *The Utility of Force: The Art of War in the Modern World* (London: Penguin, 2006).

Sottimano, A., 'The Syrian Business Elite: Patronage Networks and War Economy', *Syria Untold*, 24 September 2016, available at http://www.syriauntold. com/en/2016/09/the-syrian-business-elite-patronage-networks-and-war-economy/, accessed 16 October 2018.

Stacher, J., *Adaptable Autocrats: Regime Power in Egypt and Syria* (Stanford: Stanford University Press, 2012).

Staniland, P., 'Counterinsurgency and Violence Management' in Celeste Gventer, David Jones and M.L.R. Smith (eds), *The New Counter-Insurgency Era in Critical Perspective* (Basingstoke: Palgrave Macmillan, 2014).

Taber, R., *War of the Flea: A Study of Guerrilla Warfare Theory and Practise* (Washington DC: Brassey's, 2002).

Thompson, R., *Defeating Communist Insurgency: Experiences from Malaya and Vietnam* (London: Chatto & Windus, 1966).

Trinquier, R., *Modern Warfare: A French View of Counterinsurgency* (Westport, CN: Praeger Security International, 2006).

Ucko, D., 'The People Are Revolting': An Anatomy of Authoritarian Counterinsurgency', *Journal of Strategic Studies* 39:1 (2016).

US Army, *FM 3-24 Counterinsurgency* (Chicago: University of Chicago Press, 2006), available at http://usacac.army.mil/cac2/Repository/Materials/COIN-FM3-24.pdf, accessed 16 October 2018.

US Army, *FM 3-24 Insurgencies and Countering Insurgencies* (Chicago: University of Chicago Press, 2014), available at http://fas.org/irp/doddir/army/fm3-24.pdf, accessed 16 October 2018.

Van Dam, N., *The Struggle for Power in Syria: Politics and Society under Asad and the Ba'th Party* (London: I.B. Tauris, 2011).

Wedeen, L., *Ambiguities of Domination: Politics, Rhetoric and Symbols of Contemporary Syria* (Chicago: University of Chicago Press, 2015).

PART TWO
PRACTICAL CHALLENGES

4

THE EASY BUTTON

Foreign Military Training as an Exit Plan

Whitney Grespin

Arguably the most important military component in the War on Terror is not the fighting we do ourselves, but how well we enable and empower our partners to defend and govern themselves. The standing up and mentoring of indigenous army and police—once the province of Special Forces—is now a key mission for the military as a whole.[1]

Secretary of Defense Robert Gates, 26 November 2007

Introduction

The spread of terrorist and violent extremist organisations across borders has been accompanied by an increase in US-sponsored security assistance engagement, underpinned by a commensurate expansion of the US military's understanding of the ideological challenge it confronted. After a decade of concurrent warfare in Iraq and Afghanistan, with troop deployments of up to 160,000 and 100,000 in each country respectively, the US grew tired of large land wars fought against an enemy that used its host state's citizenry as camouflage. Driven by the idea that building partner capacity to confront threats where they originate is most effective (and a more palatable alternative to both the American public and partner

---

1 Robert Gates, *Landon Lecture by the Secretary of Defense Speech to Kansas State University on 26 November 2007* (Washington DC: US Department of Defense, 2007), available at http://archive.defense.gov/Speeches/Speech.aspx?SpeechID=1199, accessed 16 October 2018.

state[2] population than a large troop presence), this approach is also intended to reduce costs in 'lives and treasure' to the American people by reducing its overseas presence in complex environments. In unstable regions around the world, the perceived need for SFA[3] missions has grown as complex threats have continued to evolve away from state-on-state aggression and towards continuous competitions for allegiance and resources drawn from amongst the general population, in line with the 'war amongst the people' paradigm.

Given the increasing commitment to the provision of SFA as an essential component of democratisation and development, this activity merits consideration in the larger debate over the political consequences of security cooperation and its impact on civil society in recipient states. Supporting the development of professional security forces in partner countries has become a central tenet in the US fight against terrorism, violent extremist organisations, insurgencies and narco-trafficking around the globe and has elicited commensurate financial support from the Department of Defense (DoD) in such pursuits. Since the attacks of 9/11, the US Government has spent more than $250 billion developing the militaries and police forces of partners and allies.[4] Ranging from efforts to reconstruct a state military in Iraq (and in Afghanistan, the creation of an entirely new force) to attempts to help Nigeria and the Philippines to fight terrorism, the development of local security force capacity to address their own internal security has become a critical component of US military strategy.

The Obama Administration's 2015 *National Security Strategy* directed that, in addition 'to acting decisively to defeat direct threats, we will focus on building the capacity of others to prevent the causes and consequences of conflict to include countering extreme and dangerous ideologies'.[5] By helping partners combat shared threats, Washington hoped to avoid an unpalatable choice between utilising its own troops to address such

2   A state that the US works with in a specific situation or operation; in security cooperation, a state with which the Department of Defense conducts security cooperation activities. Department of Defense, *DoD Dictionary of Military and Associated Terms* (Washington DC: Department of Defense, 2018), 179, available at http://www.jcs.mil/Portals/36/Documents/Doctrine/pubs/dictionary.pdf, accessed 16 October 2018.

3   Department of Defense activities that support the development of the capacity and capability of foreign security forces and their supporting institutions. *DoD Dictionary of Military and Associated Terms*, 206.

4   Rose Jackson, *Untangling the Web: A Blueprint for Reforming American Security Sector Assistance* (Washington DC: Open Society Foundations, 2017), 6, available at https://www.opensocietyfoundations.org/reports/untangling-web-blueprint-reforming-american-security-sector-assistance, accessed 16 October 2018.

5   White House, *U.S. National Security Strategy* (Washington DC: The White House, 2015), available at http://nssarchive.us/wp-content/uploads/2015/02/2015.pdf, accessed 16 October 2018.

issues and allowing problems to grow as threats to US interests. This use of partner states as effective surrogates provides a powerful tool. As one soldier-practitioner-academic observed:

> Operationally, [SFA] is the exit strategy for costly stability operations like Afghanistan because it allows those countries to provide security for themselves rather than depend on the United States military for firepower. Strategically, helping fragile states professionalize their military and police promotes durable development, since corrupt security forces tend to devour the fruits of development.[6]

The approach, often referred to as 'small footprint', strives to be such, but does not always deliver. By being as streamlined as possible, it aims to reduce the partner state's dependence on sponsor resources, while concurrently minimising the population's awareness of (and thus potential resistance to) foreign intervention. This has a dual purpose: it avoids public knowledge of and engagement in the political drivers behind the arrangement of such military partnerships and also lessens the visibility of a US intervention, thus enhancing the standing of the recipient state by conveying the impression that the skills it acquires are 'home-grown'. Conversely, if handled poorly, such programmes risk casting the partner state as a 'puppet' of the US Government.

Developing the military capabilities of partner states is an important pillar of US national defence strategy and has occupied a central role in the DoD's campaign strategies since 9/11. It builds upon prior efforts to build partner capacity directly during conflicts such as Vietnam, as well as through longer engagements in the Americas. The post-Second World War era saw a marked expansion in security assistance,[7] security force assistance and security cooperation efforts.[8] Insofar as

---

6   Sean McFate, 'Raising an Army: Ten Rules', *War on the Rocks* July 2014, available at https://warontherocks.com/2014/07/raising-an-army-ten-rules/, accessed 16 October 2018.

7   These are defined as a group of programmes authorised by the Foreign Assistance Act of 1961 as amended; the Arms Export Control Act of 1976, as amended; or other related statutes by which the US provides defence articles, military training and other defence-related services by grant, lease, loan, credit or cash sales in furtherance of national policies and objectives, and those that are funded and authorised through the Department of State to be administered by the Department of Defense/Defense Security Cooperation Agency are considered part of security cooperation. *DoD Dictionary of Military and Associated Terms*, 206.

8   These are defined as all Department of Defense interactions with foreign security establishments to build security relationships that promote specific US security interests, develop allied and partner nation military and security capabilities for self-defence and multinational operations, and provide United States forces with peacetime and contingency access to allied and partner nations. *DoD Dictionary of Military and Associated Terms*, 206.

military advising and assistance operations have sought to provide an exit opportunity for sponsoring governments, the overall return on investment for these expenditures has been poor. It is not difficult to find examples of programme inefficiencies or procurement waste across SFA efforts, even in successful missions; it is also hard to quantify which partner capabilities have truly alleviated the operational requirements of US troops. This is due to shortcomings in monitoring and evaluation practices, inconsistent or sporadic engagement dictated by departmental limitations (troop rotations, resource scarcity, competing priorities, partner nation relationship dynamics) and the fact that, when a partner garners enough interest to receive expensive security assistance, they have likely also gained attention (and therefore investment) from other US agencies (such as the Department of State or its US Agency for International Development) who are also motivated to improve stability and governance.

This chapter examines the concept of 'war amongst the people' through the lens of building partner capacity. It commences with a discussion of the scope of contemporary US Government training of foreign militaries. Thereafter, having established a framework for the practice's implementation, it proceeds by examining the motivating factors behind the approach of 'prevention is the new victory'. This addresses the goals and merits of building partner capacity via SFA in lieu of (or complementary to) direct US military engagement. A final section offers a critique of the design of SFA efforts and an overview of lessons learned for US approaches to SFA, with a particular eye to how this practice can impact civil-military relations within the partners, as well as broader national, regional and international geopolitical concerns. It argues that, conceived and delivered with care, SFA is not only in the interests of the US, but also benefits the citizens of partner states by building a more peaceful and stable environment in which to live and work, thereby reducing the chances of 'war amongst the people' re-emerging.

Boots (and Books) on the Ground: The Current Scope of US Military-delivered SFA

SFA and Security Sector Reform (SSR)[9] focus on the way a partner state provides safety, security and justice with civilian government oversight.

---

9   This is defined as a comprehensive set of programmes and activities undertaken by a host nation to improve the way it provides safety, security and justice. *DoD Dictionary of Military and Associated Terms*, 206.

Other salient activities include Foreign Internal Defense (FID).[10] This refers to US activities that support a partner state's internal defence and development strategy and are programmes designed to protect against subversion, lawlessness, insurgency, terrorism, and other threats to their internal security and stability. These activities, which overlap, sometimes in terminology and often in practice, fall under the broader US Government effort of Defense Institution Building (DIB)[11] with partner states. The DoD's primary role in SSR is to support the reform, restructuring or reestablishment of the partner state's armed forces and the defence aspect of the security sector.

With multiple government agencies and departments engaged in these activities, it is difficult to deduce a coherent strategy and unified goals. The variety of actors and the challenge of aligning terms across doctrine and policy in a concurrent and timely fashion are two problems among many. That having been said, such efforts share a broad common goal in that they combine a desire to alleviate the need for the US to intervene directly with the hope that partnering closely enough will align strategic goals with partners who will then act in both of their interests. As the Department of State has observed, the many different forms of SFA all seek to achieve 'regional stability through effective, mutually beneficial military-to-military relations that culminate in increased understanding and defence cooperation between the US and foreign countries'.[12]

The relevance and impact of SFA activities can be inferred from the scale of US commitments. Although numbers are difficult to pin down with specificity, given the array of authorities' funding lines and differences between 'money allocated' and 'money spent', not to mention tracking challenges across fiscal years and reporting timelines, SFA is clearly a significant undertaking for the US military. Government support for security assistance to partners and allies is broad; in 2015 this included 100 different legislative authorities[13] amounting to approximately $20

---

10   This is defined as participation by civilian and military agencies of a government in any of the action programmes taken by another government or other designated organisation to free and protect its society from subversion, lawlessness, insurgency, terrorism and other threats to its security. *DoD Dictionary of Military and Associated Terms*, 93.

11   This is defined as security cooperation conducted to establish or reform the capacity and capabilities of a partner state's defence institutions at the ministerial/department, military staff and service headquarters levels. *DoD Dictionary of Military and Associated Terms*, 64.

12   Jesse Dillon Savage and Jonathan D. Caverley, *When human capital threatens the capitol: Foreign aid in the form of military training and military-backed coups* (Annapolis: US Naval War College, 2017), available at https://jessedsavage.files.wordpress.com/2010/09/jpr_final_newbib.pdf, accessed 28 October 2018.

13   For more, see Terrence K. Kelly, Jefferson P. Marquis, Cathryn Quantic Thurston, Jennifer D.P. Moroney and Charlotte Lynch, *Security Cooperation Organization in the Country*

billion.[14] In the 2017 National Defense Authorization Act, the DoD alone was slated to administer over 60 different programmes[15] to build partner capacities, with budgets totalling around half of the authority amount at roughly $10 billion.[16] The substantial budget allocations attached to US investments in building partner capacity reflect the importance attached by the government to building human capital and institutional resiliency.

In February 2018, the DoD produced an inaugural account of its broad security cooperation programmes.[17] Produced as an appendix to the President's 2019 budget submission, the document dramatically improved transparency about the Pentagon's capacity-building activities, outlining proposed funding for 27 assistance programmes totalling nearly $3.4 billion,[18] with an additional $6.6 billion of DoD money earmarked for building Afghan and Iraqi security force support in connection with ongoing military operations. However, even with these ameliorations,

---

Team: Options for Success (Washington DC: RAND, 2010), xii, available at https://www.rand.org/content/dam/rand/pubs/technical_reports/2010/RAND_TR734.sum.pdf, accessed 28 October 2018.

14 Missy Ryan, 'State Department and Pentagon Tussle Over Control of Foreign Military Aid', *The Washington Post*, 10 July 2016, available at https://www.washingtonpost.com/world/national-security/state-department-and-pentagon-tussle-over-control-of-foreign-military-aid/2016/07/10/ddc98f3e-42b0-11e6-88d0-6adee48be8bc_story.html, accessed 28 October 2018.

15 These figures do not include programmes that fall under Department of State sponsorship, which are beyond the scope of this chapter's focus on US military engagement to build capacity rather than intervene physically. This also does not include other US Government entities with an interest in building partner capacity to enhance national security. An example of this would be the Drug Enforcement Agency (DEA), which itself falls under the US Department of Justice, but has done a lot of work over the last decade in states such as Afghanistan and in Central America to interdict and mitigate drug production and trafficking. While noting the breadth of stakeholders involved in building partner capacity programmes, this chapter only looks at DoD efforts, with special attention paid to the FID efforts traditionally falling under the purview of US Special Operations Forces.

16 For details, see Office of the Under Secretary of Defense Chief Financial Officer, *Defense Budget Overview: United States Department of Defense Fiscal Year 2017 Budget Request* (Washington DC: Department of Defense, 2016), available at http://comptroller.defense.gov/Portals/45/Documents/defbudget/fy2017/FY2017_Budget_Request_Overview_Book.pdf, accessed 28 October 2018; Kathleen J. McInnis and Nathan J. Lucas, *What is 'Building Partner Capacity'? Issues for Congress* (Washington DC: Congressional Research Service, 2015), 13, available at https://fas.org/sgp/crs/natsec/R44313.pdf, accessed 28 October 2018.

17 This accounting provides a superficial picture of how resources are allocated across different regions or functions but does not include all efforts that may be related. For example, FID may be security cooperation, but security cooperation is not necessarily FID, which is a core Special Operations Forces (SOF) capability.

18 Office of the Secretary of Defense, *Fiscal Year (FY) 2019 President's Budget: Security Cooperation Consolidated Budget Display* (Washington DC: Office of the Secretary of Defense, 2018), available at http://comptroller.defense.gov/Portals/45/Documents/defbudget/fy2019/Security_Cooperation_Budget_Display_OUSDC.pdf, accessed 28 October 2018.

much remains obscured. As one analyst observed, the submission omitted 'several security cooperation programs, including the Cooperative Threat Reduction program used to reduce the proliferation of weapons of mass destruction, scholarships for foreign students attending military service academies, and international armaments cooperation'.[19]

A brief examination of the scope of the training concerned is instructive. In 2015, the number of US foreign military trainees increased substantially, growing from 56,346 the year prior to 79,865 in 2015.[20] This increase in US foreign military training was driven by a major push to improve African militaries' effectiveness in peace operations and a drive to enhance foreign security forces to better tackle the illicit drug trade and transnational organised crime groups, suggesting a continued role for focusing on 'war amongst the people' as a potential paradigm for understanding contemporary conflict. Support for Regionally Aligned Forces has focused on training programmes, joint exercises and counterterrorism efforts. From 2014–15, there was an increase in the number of trainees for all of the six major world regions except East Asia and the Pacific, which instead saw a 42 per cent drop in US trainees, despite the Obama Administration's much vaunted 'Asia Pivot'. Sub-Saharan Africa experienced the largest increase in US military trainees, closely followed by Europe and Eurasia. By including both government and contractor-delivered training, these figures give a fair impression of the scale of investment from which the US hopes to benefit when tackling these threats.

Such trends also provide some insight into the types of US military training the Trump Administration may cut as part of its effort to reduce foreign aid, which includes FID and SSR programming. The White House's proposed budget for FY2018 showed a major drop in US Foreign Military Financing (FMF), while subsequent actions taken by the Trump Administration have indicated a preference for sharing hardware and selling weapons to partners rather than continuing to focus on soft skills.[21]

---

19 Tommy Ross, 'The Pentagon Just Revealed How Much it Spends Helping Foreign Militaries', *Defense One* February 2018, available at https://www.defenseone.com/ideas/2018/02/pentagon-just-revealed-how-much-it-spends-helping-foreign-militaries/146294/?oref=defenseone_today_nl, accessed 28 October 2018.

20 Colby Goodman, Christina Arabia, Robert Watson, Taner Bertuna and Andrew Smith, 'Major Increase in US Foreign Military Training Driven by New US Peacekeeping and Counter-Narcotics Efforts' (Washington DC: Security Assistance Monitor, 2017), available at http://securityassistance.org/publication/major-increase-us-foreign-military-training-driven-new-us-peacekeeping-and-counter/, accessed 28 October 2018.

21 Mike Stone, 'Trump launches effort to boost US weapons sales abroad' *Reuters* 19 April 2018, available at https://www.reuters.com/article/us-usa-trump-arms/trump-launches-effort-to-boost-u-s-weapons-sales-abroad-idUSKBN1HIQ2E6, accessed 28 October 2018.

Trump Administration officials also called for cuts to US peacekeeping, counter-narcotics and nation-building programme budgets and have demonstrated an unwillingness to curb security assistance to countries and regimes facing allegations of serious human rights abuses.

Prevention is the New Victory

While the preceding discussion focused on *what* is being done to provide security assistance, the more important question to consider regards *why* it is being done. The idea of building partner capacity in lieu of committing itself via the deployment of combat troops to neutralise a threat has long been a practice of the DoD. Persistent peacetime engagement, military-to-military partnerships and support for diplomatic and civilian engagement programmes – all with low profiles and footprints – aim to address hybrid threats before they develop into crises. The return to persistent engagement and partner training delivery is a challenge for the US military, which has spent the last decade and a half focusing on kinetic engagements in Iraq and Afghanistan, rather than shoring up foreign military training capabilities. This is especially true of Special Operations Forces (SOF), who were frequently tasked to undertake direct action activities in lieu of applying their specialised skills to train partner militaries. A recent *SOF Operating Concept* noted the importance of returning to persistent engagement, calling for 'sustainable forward posture (forces, footprint, and agreements) in foreign countries to establish and maintain critical relations, develop and sustain critical partnerships, and support building partner capabilities and capacities [and] proficiency and sufficiency to understand and operate in the human domain'.[22] As Major Fernando Lujan wrote in a 2013 report, 'prevention is the new "victory". Instead of attempting to "surge" overwhelming resources for an elusive victory, light footprint missions aim to keep costs low, relying on a small number of civilian and military professionals to work patiently over many years to prevent and contain security challenges'.[23] After all, as any experienced climber could attest, sometimes the most advantageous place to have

22 Directorate of Force Management and Development, *Special Operations Forces Operating Concept: A Whitepaper to Guide Future Special Operations Force Development* (Washington DC: United States Special Operations Command, 2016), available at https://nsiteam.com/social/wp-content/uploads/2017/01/SOF-Operating-Concept-v1-0_020116-Final.pdf, accessed 28 October 2018.

23 Fernando M. Lujan, *Light Footprints: The Future of American Military Intervention* (Washington DC: Center for a New American Security, 2013), available at https://s3.amazonaws.com/files.cnas.org/documents/CNAS_LightFootprint_VoicesFromTheField_Lujan.pdf, accessed 28 October 2018.

your foot is a small, but solid, toehold. This makes the proposed cuts of the Trump Administration all the more significant.

One of the foremost lessons learned by the US military over the last decade of war is that a light footprint – if thoughtful and well executed – is more effective than a large presence with little guidance or precision. This logic should be applied not only to those troops that are deployed to train partner forces, but also be instilled amongst those who are to receive the training, particularly in contemporary 'wars amongst the people'. In IS-era Iraq, where then-US Defense Secretary Ash Carter estimated that recruitment for Iraqi forces had reached only 30 per cent of target numbers,[24] RAND analyst Rebecca Zimmerman advocated a different approach: 'Rather than lament the lack of force size, the Pentagon should focus instead on developing skill, discipline and the will to fight in the smaller group of Iraq's ready volunteers'.[25] Looking for the will to fight first and then developing the discipline that results in quality forces is likely to be more successful than relying on strength in numbers alone and is foundational to cohesive interrelationships amongst populations in contested environments.

As such missions typically have minimal resources and long-time horizons, they generally seek to contain or prevent security problems rather than resolve them decisively (and/or kinetically). If partners improve their governance, there is less physical space for hostile actors to exploit. As J Q Roberts of the Office of the Assistant Secretary of Defense assessed, 'if you build a net of governance it will catch all of your malign actors, or at least the great majority of them'.[26] This approach seeks to be proactive, rather than reactive. It also enforces five broad conditions that describe the stabilisation benefits that are aspired to via SFA, including the facilitation of a safe and secure environment, an established rule of law, social well-being, stable governance and a sustainable economy. This 'net of governance' is mutually reinforcing in practice, both in terms of intent and resourcing.

---

24 Deb Riechmann, 'Carter: Iraqi training goal to fall way short of recruits' *The Boston Globe* 17 June 2015, available at http://bigstory.ap.org/article/bcda0ac416be40c79f13d75814682d8f/carter-dempsey-testify-house-middle-east, accessed 28 October 2018.

25 Rebecca Zimmerman, 'Training Foreign Military Forces: Quality vs. Quantity' *War on the Rocks*, July 2015, available at http://warontherocks.com/2015/07/training-foreign-military-forces-quality-vs-quantity/, accessed 28 October 2018.

26 James Q. Roberts, 'Comments by a representative of the Office of the Assistant Secretary of Defense at the panel on Special Operative Forces and the Indirect Approach to Conflict Prevention at the 24th Annual SOL/LIC Symposium and Exhibition on *Persistent Engagement in the New Strategic Environment on* 29 January 2013' (Washington DC: National Defense Industrial Association, 2013). Further details available at https://ndia.dtic.mil/2013/SOLIC/2013SOLIC.html, accessed 28 October 2018.

Attempts to stabilise partner governments or to destabilise enemy forces is inherently a long-term institutional endeavour when taking the indirect approach. SFA is not simply handing over a weapon system with perfunctory training. As McFate has quipped, 'raising armies is more sophisticated than this, and involves engaging civil society, growing leaders, building institutions and instilling professionalism. Training and equipping alone only gives you better dressed soldiers who shoot straighter'.[27] Successful SFA missions are complex and long-term and involve activities such as reconstruction, disarmament, demobilisation and reintegration (DDR) and the development of the rule of law supported by the intervening force.[28] These capabilities go beyond the military sphere and necessitate an interagency approach to build state system-wide institutional capacity – a task that is complicated and time consuming,[29] but critical to sustaining gains. The US military saw this in both Iraq and Afghanistan, where campaigns to 'win hearts and minds', engage with local populations, identify powerbrokers and, overall, understand the human terrain of their operational environment came to feature prominently. Civilians were no longer to be avoided as a means to mitigate collateral casualties, but instead became a medium of the conflict itself, a central element of 'war amongst the people'.

Achieving the institutionalisation of gains derived from SFA training requires that policies be enacted at the military's force-generating level, to prioritise the skills, knowledge and procedures that need to be consistently present across training, education and doctrine development. Within the DoD, this is ensured through the 'Doctrine, Organization, Training, Manpower, Leadership and Education, Personnel, Facilities, and Policy' (DOTMLPF) framework. This framework is conceptually useful through its explanation of each element and has practical utility in that it mandates specific elements of capability development required from each component, detailing how such capabilities can be acquired or developed. In looking through this lens, there is a clear relationship that emerges between the operational needs and the strategic value of those skills or capabilities. This DOTMLPF framework, however, requires an appropriate level of capacity on many levels of military function, which, in its entirety, often exceeds the scope of the SFA that has been designated for programming. Building an enduring institutional capacity is the key to attaining the benefits of

---

27   McFate, 'Raising an Army: Ten Rules'.
28   Edward C. Luck, *UN Security Council: Practice and Promise* (New York: Routledge, 2006), 33.
29   Guro Lien, 'Military Advising and Assistance Operations' in Per Norheim-Martinsen and Tore Nyhamar (eds), *International Military Operations in the 21st Century* (New York: Routledge, 2015), 81.

capability generation via SFA, which requires a commitment on behalf of the partner to identify and retain quality personnel.

Furthermore, an understanding and respect for how the skills being grown by SFA training bequeath responsibility must also be imparted to recipients in the broader context of respecting the rule of law and promoting stable governance. Respect for international norms on ethics and human rights are perhaps the most universal. For example, the 'Leahy Law' used for vetting SFA recipients refers to two statutory provisions prohibiting the US Government from using funds for assistance to units of foreign security forces where there is credible information implicating that unit in the commission of gross violations of human rights, with one statutory provision applying to the State Department and the other to the DoD.[30] Accordingly, in connection with this legislation, gross violations of human rights, including incidents of torture, extrajudicial killing, enforced disappearance and rape, are examined on a 'fact-specific' basis. In this respect, the delivery of SFA by the US military has already taken a step towards certifying American confidence in the legitimacy of their partner state's military establishment, thus providing reputational capital that its own citizens might view as having value.

Ostensibly, this is the broad, aspirational goal. In the vein of changing a military culture, a 2013 report by the Center for a New American Security on the future of military operations noted that 'Professional culture and institutions cannot be changed in a single tour, but as American advisers maintain relationships with their foreign counterparts over the years, lieutenants become captains, then colonels, then generals, and they begin to influence the partner nation's military from within'.[31] This is a generational commitment when speaking about tracking and engaging over the course of an entire career, but one that remains vulnerable to changes in administration and priorities, as noted above.

Management and Lessons Learned?

Concerns regarding how the US Government conducts security assistance activities are nothing new in either Washington DC nor in field operations. In 2015, Congress recognised the importance and growth of such programmes and began an overhaul of the security cooperation enterprise

---

30 US Department of State, *Leahy Fact Sheet* (Washington DC: US Department of State, 2018), available at https://www.state.gov/j/drl/rls/fs/2018/279141.htm, accessed 28 October 2018.

31 Lujan, 'Light Footprints: The Future of American Military Intervention'.

via the drafting of the 2017 National Defense Authorization Act (NDAA).[32] In the 2017 NDAA, Congress clarified its intent to align security cooperation planning with broader US Government strategic objectives – a step that the US military would have benefited from decades before to better identify appropriate pursuits.[33]

SFA is deeply political, whether this is intentional or not. The introduction of valuable resources reconfigures the authority structure in partner states. Approaches that ignore the inherent politicisation of this endeavour will fail. Relatedly, the selection of leaders should be viewed with caution. Many militaries receiving SFA are transitional and may not have had the twenty years of leadership cultivation and professional development that the US military expects from a senior officer. In places like Iraq the military was an 'army of privates' after de-Baathification, whereas, in Afghanistan, the challenge was more about integrating non-state actors into the military and getting Northern Alliance veterans to work alongside Soviet-trained officers. Furthermore, the potential for nepotism and cronyism is rampant in circumstances with a fast influx of resources and little situational awareness – an environment that is all too familiar for SFA professionals.

There is a growing body of literature, based on the experience of the last decade and a half, regarding both the strengths and the shortcomings of private sector ability to deliver and oversee these types of training for partner militaries. [34] While a necessarily comprehensive discussion of this practice is outside the scope of this chapter, it is worth noting that private sector subject matter experts (SMEs) arguably possess more SFA expertise and responsive capability than personnel sent by the US Government and can also claim extensive recent past performance within and external to the DoD. Benefits such as capability specialisation and service modularisation, in addition to impartiality and retention of institutional memory, make contractor usage for SFA missions desirable. At the same time, they are also highly contentious, given the nature of military-to-military cooperation.

Furthermore, limitations on troop deployments or contract lengths may lead trainers to accelerate the pace of security force development and,

---

32 McInnis and Lucas, *What is 'Building Partner Capacity'? Issues for Congress.*

33 US Senate Armed Services Committee, *National Defense Authorization Act for Fiscal Year 2017* (Washington DC: US Senate Armed Services Committee, 2017), available at https://www.armed-services.senate.gov/imo/media/doc/FY17%20NDAA%20Bill%20Summary.pdf, accessed 28 October 2018.

34 For further discussion see Whitney Grespin, 'The Evolving Contingency Contracting Market: Private Sector Self-Regulation and the United States Government Monitoring of Procurement of Stability Operations Services', *PKSOI Paper* January 2016, available at http://www.dtic.mil/dtic/tr/fulltext/u2/1004033.pdf, accessed 28 October 2018.

when possible, to surge more resources than partner state institutions can handle. This can result in the promotion of corruption and a commensurate decrease in leverage, due in part to the perception that US support is unconditional and irreversible.[35] Despite funding such programmes for decades and spending billions – if not trillions – over their lifetimes, the US has struggled to achieve success. Efforts have often been very tactical in scope, focusing on 'train and equip'[36] capability generation in the short term, rather than seeking to ensure strategic gains, such as the development of sustainable capacity to replenish and build skill sets from within after the cessation of assistance. At its core, this approach to conflict prevention through increased institutional capacity is simply 'helping friends to help themselves so they can help us'.[37] SFA programmes must be complemented by robust economic, diplomatic and development engagement to advance human rights, promote sustainable development, advance the rule of law and encourage economic activity, all of which will combat the underlying drivers of instability that the US seeks to counter by enhancing partner security capabilities.

A focus on immediate capability reinforcement is understandable in states facing ongoing and concrete security threats, but this lack of focus on sustainable strategic gains means the gains from such US investments need to be demonstrated and sustained, rather than just asserted. This results in a dynamic wherein there is a need to constantly regenerate capabilities that are consumed or a need to generate capacity beyond capability. In a military context, capability encompasses the ability to perform a function to achieve a military operational objective. The Joint Staff defines capability as 'the ability to achieve a specified wartime objective,'[38] while the DoD defines a critical capability as 'a means that is considered a crucial enabler for a center of gravity[39] to function as such and is essential to the accomplishment of the specified

---

35 Lujan, 'Light Footprints: The Future of American Military Intervention'.

36 Defense Security Cooperation Agency, *Security Assistance Management Manual: C15 – Building Partner Capacity* (Washington DC: Defense Security Cooperation Agency, 2018), available at http://www.samm.dsca.mil/chapter/chapter-15, accessed 28 October 2018.

37 See Note 26.

38 Thomas Ross, 'Enhancing Security Cooperation Effectiveness: A Model for Capability Package Planning', *Joint Forces Quarterly* 80:1 (2016), 26.

39 'The center of gravity has become one of today's most popular military concepts despite the fact that its origins extend back to the early industrial-age. Clausewitz's military center of gravity (CoG) and the CoG of the mechanical sciences share many of the same properties: neither is a strength or a source of strength, per se, but rather a focal point where physical (and psychological) forces come together'. See Antulio J. Echevarria, *Clausewitz's Center of Gravity: Changing Our Warfighting Doctrine – Again!* (Carlisle: US Army War College Strategic Studies Institute, 2002), available at http://www.au.af.mil/AU/AWC/awcgate/ssi/gravity.pdf, accessed 28 October 2018.

objective(s)'.[40] As a 2016 *Joint Forces Quarterly* article explained, 'an effective military capability cannot be equated with a single weapons system; rather, it is "provided by one or more systems, and is made up of the combined effects of multiple inputs"'.[41]

While the US military has seen some success in training foreign forces to fight, there has been less success in enabling partner states to develop an indigenous force-generating function, where the goal is to impart core training knowledge (i.e. they learn how to perform tasks they could not previously perform) and to provide greater capacity (i.e. they have more forces to perform tasks vital to local and regional security).[42] It is necessary to build capacity across the strategic (executive direction), operational (force generating) and tactical (operating force) levels to ensure that the SFA is an ends-driven strategic endeavour. By changing to this approach, rather than the current means-driven efforts that currently seem to prevail, such SFA efforts would actually meet the intent of the 2017 NDAA.[43] Considering US national security interests and broader human security concerns within the context of increasingly complex operational environments – as identified in the 'war amongst the people' paradigm – demonstrates that ensuring an organic generating function capability for foreign security partners is critical to establishing and maintaining stability and deterring aggression from state threats, as well as non-state violent extremist organisations. This focus on building human capital within the partner military also encourages investments in institutional knowledge across other socio-political institutions and power structures, further strengthening armed forces' engagement with broader society at the local level.

In cases of failure, however, there may be a higher risk tied to this small footprint. Rather than taking direct military action alone, the trainer will have attempted to build capacity to accomplish an assigned mission by developing a range of military, police and irregular or specialised armed forces. However, if partner entities are unable or unwilling to defend US national security or mission interests, this may result in the need for the US Government to undertake unilateral action to supplement or intercept partner states' actions. Relatedly, there may be a need for US Government

---

40 *DoD Dictionary of Military and Associated Terms*, 56.

41 Ross, 'Enhancing Security Cooperation Effectiveness: A Model for Capability Package Planning', 26.

42 For details, see Joint Center for International Security Force Assistance, *Organizing the Generating Function of a Security Force Institution: Security, Justice, and Implications to Governance* (Fort Leavenworth: US Army War College Peacekeeping and Stability Operations Institute, 2017).

43 US Senate Armed Services Committee, *National Defense Authorization Act for Fiscal Year 2017*.

intervention in the case that the skills imparted are not applied to the intended purposes. As the author has noted elsewhere,

> It must be recognised that this sort of assistance and security force capacity-building is inherently political in that it results in an intentional allocation of resources that is determined by external stakeholders. For this reason, it is essential that capacity-building programmes take into account what unintended consequences may result from these types of capacity enhancing training schemes. Evidence of this includes unsanctioned skills transfers from former military personnel to foreign nationals on a one-to-one basis and is visible up through leadership levels.[44]

Notably, an in-depth study[45] released in May 2017 using data from 189 countries from 1970–2009 shows that the number of military officers trained by the US International Military Education and Training (IMET)[46] and Countering Terrorism Fellowship (CTFP)[47] programmes increase the probability of a military coup.[48] Interestingly, the study was published by a producer of such students itself – the US Naval War College. A key finding deduced the following:[49]

1. Approximately 61 per cent of coup attempts from 1970–2009 failed. Of those that succeeded, two-thirds were carried out by officers that had participated in IMET, leading to the conclusion that 'Successful coups are strongly associated with IMET training and spending'

2. 60 per cent of military-backed coups or coup attempts occurred in states that had received IMET training the previous year

---

44 Whitney Grespin, 'From the Ground Up: The Importance of Preserving SOF Capacity Building Skills', *Journal of Strategic Security* 7:2 (2013), 37–47, available at: http://scholarcommons.usf.edu/jss/vol7/iss2/6, accessed 28 October 2018.

45 While the US and other states provide training where the costs are covered by the receiving state, this chapter focuses only on aid or training that is not reimbursed by the recipient.

46 For details, see Defense Security Cooperation Agency, *International Military Education and Training* (Washington DC: Defense Security Cooperation Agency, 2018), available at http://www.dsca.mil/programs/international-military-education-training-, accessed 28 October 2018.

47 For details, see Defense Security Cooperation Agency, *Combating Terrorism Fellowship Program* (Washington DC: Defense Security Cooperation Agency, 2018), available at http://www.dsca.mil/programs/combating-terrorism-fellowship-program, accessed 28 October 2018.

48 Savage and Caverley, 'When human capital threatens the capitol'.

49 David Trilling, 'US-trained militaries more likely to overthrow their governments' *Journalist Resources*, August 2017, available at https://journalistsresource.org/studies/government/security-military/american-trained-militaries-overthrow-governments, accessed 28 October 2018.

3. States whose officers had participated in IMET were nearly twice as likely to experience a coup or coup attempt

4. When the number of officers trained (or the amount the US spent) moves from the 25th percentile to the 75th percentile, the chance of a coup or coup attempt doubles. IMET is a subset of US SFA. In FY 2015, the US spent a total of $876.5 million to train over 76,000 soldiers from 154 countries; about $300 million of that was aid (costs not covered/reimbursed by the recipient country)

5. US military aid (weapons, for example, but excluding training) has a slightly negative relationship to coup probability, supporting the hypothesis that provision of weapons enables regimes to protect or 'coup-proof' themselves

6. Other foreign aid, such as food, for example, has no discernible statistical relationship to coup probability.

In short, the authors concluded that 'training imparts valuable resources to and increases the professional distance of a potentially dangerous section of a developing state's polity [the military]. Increasing trainees' human capital is likely to increase resource demands on the regime and improve the military's ability to remove the regime should its demands not be met'.[50]

In addition to the risk of the training being used against its own government, a further concern looms: the SFA may simply be unable to prevent a state from failing – an outcome that defeats the very purpose of the programmes in question. When facing failure, this 'engagement as an exit strategy' can metastasise. In the case of an impending fall of a US-supported regime, there is likely to be an evolving imperative to transition from a FID campaign to an unconventional warfare campaign if the possibility is assessed as a viable strategic option to reinstall a US-friendly regime.[51] Case studies such as Cuba, Iran and Nicaragua are demonstrative of when the US abandoned foreign military training as an exit strategy and instead faced decisions about the viability of transferring efforts to unconventional warfare to replace opposition regimes.

This notion of transitioning from SFA to unconventional warfare is based on the US experience in Vietnam, during which time FID emerged as a doctrinal concept and necessity to shore up the viability of a non-communist South Vietnamese regime. The production of the 1963 version

---

50   Savage and Caverley, 'When human capital threatens the capitol'.

51   Jason Martinez, 'From Foreign Internal Defense to Unconventional Warfare: Campaign Transitions when US-Support to Friendly Governments Fails'. Thesis submitted to the US Army Command and General Staff College in May 2015 (Fort Leavenworth: US Army Command and General Staff College, 2015), available at http://www.dtic.mil/dtic/tr/fulltext/u2/1000485.pdf, accessed 28 October 2018.

of *Field Manual (FM) 31-22, U.S. Army Counterinsurgency Forces* focused on training, advising and assisting indigenous forces[52] – an effort that has come back into vogue through Train, Advise, Assist (TAA) mandates for many overseas missions. In Vietnam, the US-sponsored SFA campaign escalated the ferocity and breadth of military engagement across the region. However, after the fall of Saigon and the failure of the overall mission, the DoD took a step back from such efforts until necessity required similar reengagement elsewhere.

Assisting with the professionalisation of both security forces and broader state administrative capacity promotes the necessary durable development that will allow for gains to be held. By mitigating instability where it originates – out of populations lacking participatory opportunities within their own communities and faith in their own government's ability to provide physical and socioeconomic stability – capacity-building of state institutions developed concurrently with SFA can serve to positively engage the population.

From the 1950s through the 1970s, the US provided support to partner states with varying levels of success, but it was in the 1980s that the DoD saw what it perceived as success via SFA to defeat an insurgency in El Salvador and, through the provision of unconventional warfare support to those opposing Soviet rule in Afghanistan, to eventually cause the fall of the Moscow satellite regime. However, this use of SFA to non-state actors via support of unconventional warfare ultimately failed because the US Government did not commit to supporting the development of a legitimate government that sought to promote self-sustaining stability and security. This is just one of many examples that demonstrate that comprehensive SFA – primarily via the development of a legitimate rule of law – is critical and that simply providing military materiel and training do not a viable state make. Clearly, this has been a hard lesson to learn, as recent losses in Iraq and Afghanistan have demonstrated.

In the case of imminent SFA failure – arguably something the DoD has faced recently or is facing in Afghanistan, Iraq and Yemen – determinations must be made whether there will be a withdrawal or a transition to unconventional warfare. These determinations dictate what type of preparation of the environment should be undertaken while the state remains at least semi-permissive, prior to the fall of the US-supported

---

52 Department of the Army, *Field Manual 31-22, U.S. Army Counterinsurgency Force* (Washington DC: US Government Printing Office, 1963), available at https://www. scribd.com/document/328189453/U-S-ARMY-COUNTERINSURGENCY-FORCES-DEPARTMENT-OF-THE-ARMY-FIELD-MANUAL-FM-31-22-NOVEMBER-1963, accessed 28 October 2018.

government.[53] All this being said, the relationships and networks that these light-footprint missions produce represent a small, but important, investment to hedge against future 'black swan' contingencies.[54] Even when these preventive measures fail and tensions escalate into armed conflict, a small footprint can still result in some success, even if this is limited to avenues of relationship-based communication and negotiation.

Conclusion

This chapter has argued that the US Government should not fall into the trap of what Samuel Huntington identified as 'strategic monism'.[55] This occurs when one regards a specific geographic environment, threat or capability as the main driver of strategy – in this case, foreign military training as an exit strategy. The idea that the US can, and should, focus on one superior capability has shifted from SOF direct action capabilities to capacity-building over the course of the decade and a half since 9/11. This approach is based on the idea that it is possible to predict the characteristics of future warfare – such as the continuation of 'war amongst the people' – and then shape those future conflicts to maximise the utility of a specialised capability. The nature of warfare since the fall of the Berlin Wall, as well as its shift from great power competition to war's prosecution 'amongst the people', has showed the international community that these predictions are difficult to make and that unanticipated factors, such as technological revolutions (in the case of social media), can lead to ideological ones (such as the Arab Spring) that pose challenges and opportunities unlike those seen before.

While well practised, using SFA to minimise US military boots on the ground risks making every problem look like a nail when the only weapon or skillset is a hammer. Instead of investing more dollars in diplomacy and development to mitigate insecurity, its continued use socialises the idea that defence via SFA is not only viable but is also sustainable. Furthermore, it gives rise to the question of whether the practice is, in fact, an 'easy button' that allows the US military to have a visible (if small) footprint, allowing them to claim that they are contributing to stability and capacity, but whether they are tied to actual, standard sustainable metrics is another story. If the US Government continues to train its partners using short-term

---

53 Martinez, 'From Foreign Internal Defense to Unconventional Warfare'.
54 For details, see Nassim Taleb, *The Black Swan: The Impact of the Highly Improbable* (New York: Random House, 2007).
55 See Samuel P. Huntington, *The Soldier and the State: The Theory and the Politics of Civil Military Relations* (Harvard: Harvard University Press, 1957).

solutions, via the provision of military tactical capacity enhancement, at the expense of operational and strategic capability development paired with holistic approaches to reinforcing the 'net of governance' and its core conditions, it will never achieve long-term success in its goal of confronting threats at source.

References

Defense Security Cooperation Agency, *Combating Terrorism Fellowship Program* (Washington DC: Defense Security Cooperation Agency, 2018), available at http://www.dsca.mil/programs/combating-terrorism-fellowship-program, accessed 28 October 2018.

Defense Security Cooperation Agency, *International Military Education and Training* (Washington DC: Defense Security Cooperation Agency, 2018), available at http://www.dsca.mil/programs/international-military-education-training-, accessed 28 October 2018.

Defense Security Cooperation Agency, *Security Assistance Management Manual: C15 – Building Partner Capacity* (Washington DC: Defense Security Cooperation Agency, 2018), available at http://www.samm.dsca.mil/chapter/chapter-15, accessed 28 October 2018.

Department of Defense, *DoD Dictionary of Military and Associated Terms* (Washington DC: Department of Defense, 2018), available at http://www.jcs.mil/Portals/36/Documents/Doctrine/pubs/dictionary.pdf, accessed 16 October 2018.

Directorate of Force Management and Development, *Special Operations Forces Operating Concept: A Whitepaper to Guide Future Special Operations Force Development* (Washington DC: United States Special Operations Command, February 2016), available at https://nsiteam.com/social/wp-content/uploads/2017/01/SOF-Operating-Concept-v1-0_020116-Final.pdf, accessed 28 October 2018.

Echevarria, A.J., *Clausewitz's Center of Gravity: Changing Our Warfighting Doctrine – Again!* (Carlisle: US Army War College Strategic Studies Institute, 2002), available at http://www.au.af.mil/AU/AWC/awcgate/ssi/gravity.pdf, accessed 28 October 2018.

Goodman, C., Arabia, C., Watson, R., Bertuna, T. and Smith, A., 'Major Increase in U.S. Foreign Military Training Driven by New U.S. Peacekeeping and Counter-Narcotics Efforts' (Washington DC: Security Assistance Monitor, May 2017), available at http://securityassistance.org/publication/major-increase-us-foreign-military-training-driven-new-us-peacekeeping-and-counter/, accessed 28 October 2018.

Grespin, W., 'From the Ground Up: The Importance of Preserving SOF Capacity Building Skills', *Journal of Strategic Security* 7:2 (2013), available at: http://scholarcommons.usf.edu/jss/vol7/iss2/6, accessed 28 October 2018.

Grespin, W., 'The Evolving Contingency Contracting Market: Private Sector Self-Regulation and the U.S. Government Monitoring of Procurement of Stability Operations Services', *PKSOI Paper* 2016, available at http://www.dtic.mil/dtic/tr/fulltext/u2/1004033.pdf, accessed 28 October 2018.

Huntington, S.P., *The Soldier and the State: The Theory and the Politics of Civil Military Relations* (Harvard: Harvard University Press, 1957).

Jackson, R., *Untangling the Web: A Blueprint for Reforming American Security Sector Assistance* (Washington DC: Open Society Foundations, January 2017), available at https://www.opensocietyfoundations.org/reports/untangling-web-blueprint-reforming-american-security-sector-assistance, accessed 16 October 2018.

Joint Center for International Security Force Assistance, *Organizing the Generating Function of a Security Force Institution: Security, Justice, and Implications to Governance* (Fort Leavenworth: US Army War College Peacekeeping and Stability Operations Institute, 2017).

Kelly, T., Marquis, J., Thurston, C., Moroney, J. and Lynch, C., *Security Cooperation Organization in the Country Team: Options for Success* (Washington DC: RAND, 2010), available at https://www.rand.org/content/dam/rand/pubs/technical_reports/2010/RAND_TR734.sum.pdf, accessed 28 October 2018.

Lien, G. 'Military Advising and Assistance Operations' in Per Norheim-Martinsen and Tore Nyhamar (eds), *International Military Operations in the 21st Century* (New York: Routledge, 2015).

Luck, E.C. *UN Security Council: Practice and Promise* (New York: Routledge, 2006).

Lujan, F.M., *Light Footprints: The Future of American Military Intervention* (Washington DC: Center for a New American Security, 2013), available at https://s3.amazonaws.com/files.cnas.org/documents/CNAS_LightFootprint_VoicesFromTheField_Lujan.pdf, accessed 28 October 2018.

McFate, S., 'Raising an Army: Ten Rules', *War on the Rocks*, July 2014, available at https://warontherocks.com/2014/07/raising-an-army-ten-rules/, accessed, 16 October 2018.

McInnis, K.J., and Lucas, N.J., *What is 'Building Partner Capacity'? Issues for Congress* (Washington DC: Congressional Research Service), 18 December 2015, available at https://fas.org/sgp/crs/natsec/R44313.pdf, accessed 28 October 2018.

Martinez, J., 'From Foreign Internal Defense to Unconventional Warfare: Campaign Transitions when US-Support to Friendly Governments Fails'. Thesis submitted to the US Army Command and General Staff College in May 2015 (Fort Leavenworth: US Army Command and General Staff College, 2015), available at http://www.dtic.mil/dtic/tr/fulltext/u2/1000485.pdf, accessed 28 October 2018.

Norheim-Martinsen, P. and Nyhamar, T. (eds), *International Military Operations in the 21st Century* (New York: Routledge, 2015).

Office of the Under Secretary of Defense Chief Financial Officer, *Defense Budget Overview: United States Department of Defense Fiscal Year 2017 Budget Request* (Washington DC: Department of Defense, 2016).

Ross, T. 'Enhancing Security Cooperation Effectiveness: A Model for Capability Package Planning', *Joint Forces Quarterly* 80:1 (2016).

Ross, T., 'The Pentagon Just Revealed How Much it Spends Helping Foreign Militaries', *Defense One*, February 2018, available at https://www.defenseone.com/ideas/2018/02/pentagon-just-revealed-how-much-it-spends-helping-foreign-militaries/146294/?oref=defenseone_today_nl, accessed 28 October 2018.

Ryan, M., 'State Department and Pentagon Tussle Over Control of Foreign Military Aid', *The Washington Post*, 10 July 2016, available at https://www.washingtonpost.com/world/national-security/state-department-and-pentagon-tussle-over-control-of-foreign-military-aid/2016/07/10/ddc98f3e-42b0-11e6-88d0-6adee48be8bc_story.html, accessed 28 October 2018.

Savage, J.D. and Caverley, J.D., *When human capital threatens the capitol: Foreign aid in the form of military training and military-backed coups* (Annapolis: US Naval War College, 2017), available at https://jessedsavage.files.wordpress.com/2010/09/jpr_final_newbib.pdf, accessed 28 October 2018.

Taleb, N. *The Black Swan: The Impact of the Highly Improbable* (New York: Random House, 2007).

Trilling, D., 'US-trained militaries more likely to overthrow their governments', *Journalist Resources*, 30 August 2017, available at https://journalistsresource.org/studies/government/security-military/american-trained-militaries-overthrow-governments, accessed 28 October 2018.

US Senate Armed Services Committee, *National Defense Authorization Act for Fiscal Year 2017* (Washington DC: US Senate Armed Services Committee, 2017), available at https://www.armed-services.senate.gov/imo/media/doc/FY17%20NDAA%20Bill%20Summary.pdf, accessed 28 October 2018.

White House, *U.S. National Security Strategy* (Washington DC: The White House, 2015), available at http://nssarchive.us/wp-content/uploads/2015/02/2015.pdf, accessed 16 October 2018.

Zimmerman, R., 'Training Foreign Military Forces: Quality vs. Quantity', *War on the Rocks*, July 2015, available at http://warontherocks.com/2015/07/training-foreign-military-forces-quality-vs-quantity/, accessed 28 October 2018.

5

CONCEPTUALISING THE REGULAR–IRREGULAR ENGAGEMENT

The Strategic Value of Proxies and Auxiliaries in 'Wars Amongst the People'

Vladimir Rauta

Introduction

The notion of 'war amongst the people' is a central feature of the twenty-first century security environment. Introduced by Rupert Smith in his ground-breaking *The Utility of Force*,[1] 'war amongst the people' captured a reality long in the making, whose historical lineage could partly be traced back to the origins of war itself. The appeal of the concept came from combining the simplicity of the label with its strong analytical power. Smith shifted the strategic mindset towards the socio-political construction of violence in a way that allowed Western strategic thinking to grasp realities that did not conform to mainstream strategic expectations: first, the transformation of civil society into a battlespace dominated by fragmented non-state actors pursuing various and often contradictory political goals; second, the blurring of key strategic conceptual binaries such as 'peace–war' or 'victory–defeat'; and third, the increasing media visibility of such development and interactions and its taxing pressures on policy and decision-makers.

---

1 Rupert Smith, *The Utility of Force: The Art of War in the Modern World* (London: Penguin, 2006).

In doing so, Smith identified 'the people' as the *locus* and *animus* of fighting and made the case for the absence of any form of boundaries around them, physical or not. More recently, Thomas Marks and Paul Rich described the value of violence in 'war amongst the people' as a twofold process: 'to carry out the normal functions of military warfighting, neutralisation of the armed capacity of the enemy; but, more fundamentally, to carve out the space necessary for the political activities of (alternative) state-building achieved through mobilisation and construction of capacity'.[2] This shows how 'the people' became the object of contention or what needed to be won over, while Western strategic thinking adopted the famous 'hearts and minds' model to varying degrees of success.

This chapter addresses a significant gap in this debate by looking at how these wars are often fought *against* an adversary; not just *amongst* the people, but *with and alongside* the people. To highlight the intricacies of 'war amongst the people', it identifies two complementary strategic models of integrating 'the people' into warfighting: the auxiliary strategic model and the proxy strategic model, both of which speak to different patterns of interaction between regular and irregular forces. The former delineates a close regular–irregular military synergy in which the irregulars complement the regulars and are usually co-employed in the fighting. The latter describes a strategic relationship of political *and* military value in which the irregulars work for the regulars through a process of delegation. The chapter, therefore, builds a case for differencing proxies from auxiliaries, based on the former's *politico-strategic role* compared to the latter's *military-tactical utility*. To capture these differences, the argument presents the proxy and auxiliary relationships as variations of dynamic and flexible strategic interaction processes between types of forces (regular and irregular).

Historically, both models demonstrate strategic appeal. The auxiliary model can be traced back to the seventeenth century, continuing into the nineteenth century with partisans acting in concert with emerging European armies and in many wars of colonial domination.[3] The proxy model also has a rich historical tradition that reaches fruition with the Cold War and its ensuing superpower confrontation, which plunged the so-called Third World into the hot wars of the era.[4] Both models survived

---

2   Thomas Marks and Paul Rich, 'Back to the Future: People's War in the 21st Century', *Small Wars and Insurgencies* 28:3 (2017), 409–25.

3   Beatrice Heuser (ed.), *Small Wars and Insurgencies in Theory and Practice, 1500–1850* (London and New York: Routledge, 2017).

4   See Geraint Hughes, *My Enemy's Enemy: Proxy Warfare in International Politics* (Eastbourne: Sussex Academic Press, 2012); Michael A. Innes (ed.), *Making Sense of Proxy Wars: States, Surrogates and the Use of Force* (Washington DC: Potomac Books, 2012); Andrew Mumford, *Proxy Warfare* (London: Polity, 2013); Seyom Brown, 'Purposes and Pitfalls of War by Proxy:

the post-Cold War security environment and have become a staple of recent military adventurism in the Middle East and South and Central Asia. However, their strategic appeal (even less the strategic differences between auxiliaries and proxies) are seldom discussed comparatively, if at all.[5] As Scheipers observed about auxiliaries, 'the failure of Western officers and strategic thinkers to engage in a debate over the strategic value of native auxiliaries is puzzling, given the ubiquity with which local auxiliaries were – and continue to be – used'.[6] This chapter addresses this gap and uses Afghanistan's history of war as a theory-building case study.

Afghanistan's war history is instructive in multiple ways. First, through its long history of military interventions and civil war, it has become a paradigmatic case of 'war amongst the people'. From the nineteenth century British and Russian imperialist interventions to the decade-long Soviet agony prefacing the end of the Cold War, through to the post-9/11 American interventionist failure, Afghanistan has become synonymous with complicated violent people's struggles leading to the impossibility of success and the certainty of defeat. The Taliban conform to Smith's prototype of insurgents who integrate the 'people' in a complex manner and by blurring the civilian-military distinction. While hierarchically structured, the movement comprises various networks, such as that led by Sirajuddin Haqqani, or integrates a complicated web of regional and provincial tribes, such as those of Baz Mohammed and Mansoor Dadullah.[7] More importantly, current conflict resolution approaches have favoured peace deals and a welcoming of the Taliban to the negotiating table. These developments come against the background of a recent Taliban resurgence,[8] which has legitimised the group in the on-going peace talks.[9] As Afghan

---

A Systemic Analysis', *Small Wars and Insurgencies* 27:2 (2016), 243–57; Andreas Krieg, 'Externalizing the Burden of War: the Obama Doctrine and US Foreign Policy in the Middle East', *International Affairs* 92:1 (2016), 97–113; Alex Marshall, 'From Civil War to Proxy War: Past History and Current Dilemmas', *Small Wars and Insurgencies* 27:2 (2016), 183–95.

5   Vladimir Rauta, 'Proxy Agents, Auxiliary Forces, and Sovereign Defection: Assessing the Outcomes of Using Non-State Actors in Civil Conflicts', *Southeast European and Black Sea Studies* 16:1 (2016), 91–111.

6   Sibylle Scheipers 'Counterinsurgency or Irregular Warfare? Historiography and the Study of Small Wars', *Small Wars and Insurgencies* 25:5/6 (2014), 879–99.

7   Theo Farrell and Michael Semple, 'Making Peace with the Taliban', *Survival* 57:6 (2015), 79–110.

8   Reuters, 'Afghan Taliban Launch Spring Offensive as US Reviews Strategy', *Reuters*, 28 April 2017, available at https://www.reuters.com/article/us-afghanistan-taliban/afghan-taliban-launch-spring-offensive-as-u-s-reviews-strategy-idUSKBN17U0E9, accessed 17 October 2018.

9   Hekmat Khalil Karzai, 'An Unprecedented Peace Offer to the Taliban', *New York Times* 12 March 2018, https://www.nytimes.com/2018/03/11/opinion/peace-taliban.html, accessed 17 October 2018.

President Ashraf Ghani observed during 2018, it is the Afghan people who demanded peace in the hope that decades of 'war amongst the Afghan people' make way for decades of future peace.[10]

Second, Afghanistan's historical trajectory has been widely presented as having been shaped by regional and international geopolitical struggles. This gave rise to the mainstream argument that Afghanistan devolved from a somewhat sovereign buffer state into a war-torn proto-state, manned by warlords and violent factions that were always willing to barter the future of the country. This has translated into analyses of Afghan violence that use the labels 'proxy' and 'auxiliary' in an interchangeable fashion and with significant analytical consequences. This is even more puzzling as Afghanistan has been subject to a vast array of scholarship that has analysed its violence on macro, meso and micro levels. As will be demonstrated, the employment of local forces as either auxiliary or proxies[11] and the disregard of their core differences led to significant setbacks on the battlefield, be it political or military. Not every local force or militia working *with* or *for* the regular forces was always a proxy, nor was it always an auxiliary. The two models of regular–irregular interaction co-existed and, by providing a trans-historic analysis, the chapter seeks to theorise this often-recurring problem.[12]

Overall the chapter seeks to present the proxy-auxiliary issue as part of the broader narrative of 'war amongst the people'. By underlying their fundamentally strategic differences, the argument developed here tentatively helps to overcome problems surrounding issues such as counterinsurgency and counterterrorism,[13] or democratisation and reconstruction,[14] all of which have revolved around the centrality of the

---

10 Ashraf Ghani, 'I will negotiate with the Taliban anywhere', *The New York Times*, 27 June 2018, available at https://www.nytimes.com/2018/06/27/opinion/ashraf-ghani-afghanistan-president-peace-talks-taliban-.html, accessed on 17 October 2018.

11 Barnett R. Rubin, 'Women and Pipelines. Afghanistan's Proxy Wars', *International Affairs* 73:2 (1997), 283–96.

12 Stephen Biddle, 'Afghanistan's Legacy: Emerging Lessons of an Ongoing War', *The Washington Quarterly* 37:2 (2014), 73–86.

13 David Betz and Anthony Cormack, 'Iraq, Afghanistan and British Strategy', *Orbis* 53:2 (2009), 319–36; Rudra Chaudhuri and Theo Farrell, 'Campaign disconnect: operational progress and strategic obstacles in Afghanistan, 2009–2011', *International Affairs* 87:2 (2011), 271–96; Robert Egnell, 'Lessons from Helmand, Afghanistan: what now for British counterinsurgency?', *International Affairs* 87:2 (2011), 297–315; Stuart Griffin, 'Iraq, Afghanistan and the future of British military doctrine: from counterinsurgency to stabilization', *International Affairs* 87:2 (2011), 317–33; Theo Farrell and Antonio Giustozzi, 'The Taliban at war: inside the Helmand insurgency, 2004–2012', *International Affairs* 89:4 (2013), 845–71.

14 Peter Marsden, 'Afghanistan: the Reconstruction Process', *International Affairs* 79:1 (2003), 91–105; Barnett R. Rubin, 'Transnational Justice and Human Rights in Afghanistan', *International Affairs* 79:3 (2003), 567–81; Jan Angstrom, 'Inviting the Leviathan: external

'people'. Finally, given the recent advent in the practice of proxy wars, with wars such as those in Syria, Ukraine, Yemen or South Sudan,[15] and considering the growing use of auxiliary forces in recent counterinsurgency campaigns, such as Afghanistan, Iraq, and Syria,[16] the chapter furthers the understanding of the strategic differences between the two roles – proxy and auxiliary – as part of the volume's aim to conceptualise and clarify the ever-present puzzles of contemporary 'war amongst the people'.

The chapter unfolds in two parts. First, it presents the theoretical argument. Here the focus is on drawing a theoretical demarcation line between the auxiliary and the proxy model by employing the strategic interaction framework. Simply put, strategic interaction refers to a decision-making process in which one actor's options and decisions are taken in relationship with another's alternatives and commitments.[17] The choice of this framework is not incidental, but rather speaks to the core of Smith's conceptualisation of 'war amongst the people' as essentially complex political processes. A second substantive section uses Afghanistan as a theory-building case and tracks the historical evolution of the proxy and auxiliary strategic models.[18]

The Missing Link: Strategy, Proxies and Auxiliaries

The attempt to use the strategic interaction framework in order to explain variation in the employment of irregulars in operations with regular forces follows a recent, albeit slow, turn in conflict research.[19] More widely, however, it responds to both a call for abandoning non-

---

forces, war, and state-building in Afghanistan', *Small Wars and Insurgencies* 19:3 (2008), 374–96; Toby Dodge, 'Intervention and dreams of exogenous statebuilding: the application of Liberal Peacebuilding in Afghanistan and Iraq', *Review of International Studies* 39 (2013), 1189–1212; David Romano, Brian Calfano and Robert Phelps, 'Successful and Less Successful Interventions: Stabilizing Iraq and Afghanistan', *International Studies Perspectives* 16 (2015), 388–405.

15 Vladimir Rauta and Andrew Mumford, 'Proxy Wars and the Contemporary Security Environment' in Robert Dover, Huw Dylan and Michael S. Goodman (eds), *The Palgrave Handbook of Security, Risk and Intelligence* (London: Palgrave Macmillan, 2017), 99–116.

16 Kevin Koehler, Dorothy Ohl and Holger Albrecht, 'From Disaffection to Desertion: How Networks Facilitate Military Insubordination in Civil Conflict', *Comparative Politics* 48:4 (2016), 439–57.

17 David A. Lake and Robert Powell (eds), *Strategic Choice and International Relations* (Princeton: Princeton University Press, 1993), 3.

18 Tim Bird and Alex Marshall, *Afghanistan: How the West Lost Its Way* (Yale: Yale University Press, 2011); Frank Ledwidge, *Losing Small Wars: British Military Failure in the 9/11 Wars* (Yale: Yale University Press, 2017).

19 Belgin San-Akca, *States in Disguise: Causes of State Support for Rebel Groups* (Oxford: Oxford University Press, 2016).

strategic analyses of wars[20] and the need to think creatively about political violence in contemporary conflicts.[21] To assess the proxy-auxiliary difference through the idea of 'strategy' might seem futile, given the latter's controversial nature. Strachan famously decried the loss of meaning of 'strategy' and its ever-growing banal use,[22] while Freedman postulated, at the very beginning of his study, *Strategy: A History*, that 'there is no agreed-upon definition of strategy that describes the field and limits its boundaries'.[23] Without bypassing the importance of this debate,[24] the chapter employs Betts' definition of strategy as 'the link between military means and political means, the scheme for how to make one produce the other'.[25] What is relevant from the notion of 'strategy' is its ability to translate actor behaviour in a dynamic way. It simply does not assume it *ex ante*, but allows for intent to be constructed through interactions: with one's goals and means, with one's targets, with the targets' goals and means, as well as with the context and operational environment. Strategy serves, therefore, because it is fundamentally relational, hence why the chapter draws on the theoretical value of 'strategic interaction'. As Lake and Powell put it, strategic interaction refers to 'each actor's ability to further its ends depends on how others behave, and, therefore each actor must take the actions of others into account'.[26]

This framing of the proxy and auxiliary models helps to overcome difficulties arising from the often messy and covert processes though which irregulars assume these roles. This is achieved because, with strategic interaction, the focus is on how parties act, react, anticipate, presume or negate behaviour in relations to other actors and to the context. Critically, it links two problems: first, *who* is involved, namely the regular and irregular actors and second, *why* and *how* they interact. As Cunningham, Gleditsch, and Salehyan argued, the failure to specify who fights hinders the discussion on why they do it.[27]

---

20 David E. Cunningham, Kristian Skrede Gleditsch and Idean Salehyan, 'It Takes Two: A Dyadic Analysis of Civil War Duration and Outcome', *Journal of Conflict Resolution* 53:4 (2009), 570–97.

21 Paul Staniland, 'States, Insurgents, and Wartime Political Orders', *Perspectives on Politics* 10:2 (2012), 243–64.

22 Hew Strachan, 'The Lost Meaning of Strategy', *Survival* 47:3 (2005), 34.

23 Lawrence Freedman, *Strategy: A History* (Oxford: Oxford University Press, 2013), xi.

24 For a recent overview of the debate see Paul D. Miller, 'On Strategy, Grand and Mundane', *Orbis* 60:2 (2016), 237–47.

25 Richard K. Betts, 'Is Strategy an Illusion?', *International Security* 25:2 (2000), 5–50.

26 Lake and Powell, *Strategic Choice and International Relations*.

27 Cunningham, Gleditsch and Salehyan, 'It Takes Two: A Dyadic Analysis of Civil War Duration and Outcome', 571.

This is critical to our understanding of the 'people' component in the overall concept of 'war amongst the people', particularly if one views the 'people' as more than simply the object of a 'hearts and minds' campaign. Smith's formulation of the notion aimed to capture the political agency of the many non-state actors and their tremendous ability to segment the political space in ways that states did not. From this point of view, a strategic analysis of how the 'people' contribute to fighting brings to the forefront the very issue of the agency of the 'people'. Accordingly, it is the strategic intent behind both actors' behaviour and their individual goals that shape the choice of proxy and auxiliary, as well as their willingness to assume strategic responsibilities.

Having explained the choice of theoretical framework and why strategic interaction works to explain the differences between the proxy and the auxiliary models, it is possible to conceptualise the two relationships thus: proxy forces serve a *politico-strategic* role, whereas auxiliaries present a *military-tactical* value. This is consistent with the limited attempts in the literature to distinguish between the two types of force.[28] On the one hand, auxiliaries have been defined as 'military forces that support the military efforts of regular armed forces of a state',[29] while, on the other, proxy forces have been defined through the wider phenomenon, namely proxy war. This has been defined as 'the indirect engagement in a conflict by third parties wishing to influence its strategic outcome'.[30] The definition is also complemented by what can be called a structural definition, one that presents the unique structuring of a proxy war as a relationship between 'a benefactor, who is a state or non-state actor external to the dynamic of an existing conflict, and their chosen proxies, who are the conduit for weapons, training and funding from the benefactor'.[31]

To better underline the fact that proxy forces serve a *politico-strategic* role (whereas auxiliaries present a *military-tactical* value), the chapter draws two modes of interaction that demonstrate the differences in employing proxies and auxiliaries, by showing how they either *conserve* or *modify* the number of parties involved in fighting [see Figures 5.1 and 5.2]. The emphasis here is on *how* the parties interact and not on *why* they engage in such roles or, for that matter, *to what end*. The literature provides some answers,[32] yet a full discussion of the questions of why and to what end

28   Sibylle Scheipers, *Unlawful Combatants: A Genealogy of the Irregular Fighter* (Oxford: Oxford University Press, 2015).

29   Sibylle Scheipers, 'Irregular Auxiliaries after 1945', *The International History Review* 39:1 (2017), 14–29.

30   Mumford, *Proxy Warfare*, 1.

31   Mumford, *Proxy Warfare*, 11.

32   Mumford, *Proxy Warfare*, 13.

exceeds the limits of this chapter. For example, research has shown that a proxy war is the result of a colluding effort,[33] understood as a form of covert delegation of violence 'often entailing specific cooperative modalities'.[34] In terms of their purpose, the literature has used the case of the Russian annexation of Crimea to note that auxiliaries played 'the role of justifying and legitimizing the intervention with their actions being portrayed as supportive to the covert military intervention'.[35]

With a focus on *how* the regular-irregulars interact, it is necessary to distinguish between proxy forces – in which the irregulars fight the adversary *for* the regular forces – and auxiliary forces, where the irregulars fight the adversary *with and alongside* the regulars. In the case of the proxy forces, the fighting dynamic is altered because the fighting between the regulars and their targets is shifted onto the proxy. This is why proxy forces modify the number of parties involved in fighting, effectively shifting the burden of war. Conversely, in the case of auxiliary forces, these do not change the nature of who engages the adversary because they act as force multipliers – in the same or in a different theatre – for the regulars. That is why they conserve the number of parties involved in fighting. An auxiliary becomes part of a direct, overt alliance where the effort of the third party is cooperatively integrated into that of the party requesting it. There are many historical examples of campaigns involving auxiliaries: tribal chiefs working with the Zimbabwe African People's Union (ZAPU) against Rhodesia/Zimbabwe; the Mau Mau Kikuyu auxiliaries helping the British Army during the Mau Mau uprising in Kenya; the Tropas Nomadas assisting the Spanish in Western Sahara and the tirailleur regiments, the Moghnaznis, or the Harkis fighting alongside the French in Algeria.

What is particular to the proxy model is that the relationship between the Beneficiary, Proxy Actor and Target results in an overlap of three interactions: the Beneficiary-Target (the lighter circle in Figure 5.1), the Beneficiary-Proxy Agent (the semi-dotted circle in Figure 5.1) and the Proxy Agent-Target (the darker circle in Figure 5.1). The specificity of the proxy model is that it amounts to a proxy war: the indirect projection of violence onto the Beneficiary-Target interaction via the Proxy Agent-Target interaction through the Beneficiary-Proxy Agent interaction. This is different to the employment of an auxiliary, which does not result in the formation of a distinct war but marks direct, cooperative strategic

---

33 Paul Staniland, 'Armed Groups and Militarized Elections', *International Studies Quarterly* 59 (2015), 694–705.

34 Zeev Maoz and Belgin San-Akca, 'Rivalry and State Support of Non-State Armed Groups (NAGs), 1946–2001', *International Studies Quarterly* 56 (2012), 721.

35 Rauta, 'Proxy Agents, Auxiliary Forces, and Sovereign Defection'.

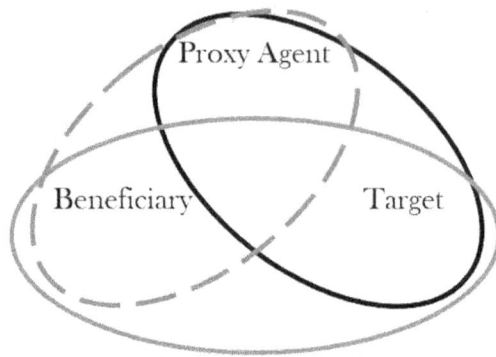

Figure 5.1   Party Interaction in Proxy Wars

Source: Vladimir Rauta

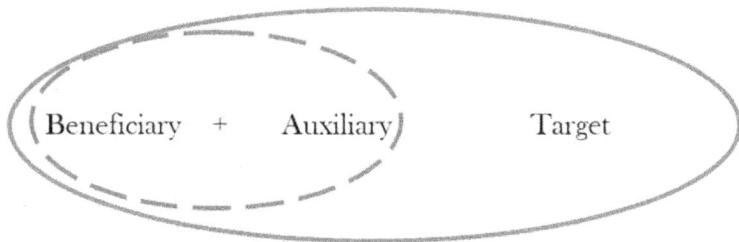

Figure 5.2   Party Interaction with Auxiliary Employment

Source: Vladimir Rauta

behaviour. In fact, throughout history, the degree of collaboration with the auxiliaries saw their quasi-assimilation into army ranks, such as the role of Native Americans before and during the American War of Independence (1775–83). Here, the combination of European conventional forces with unconventional auxiliaries was common; George Washington's success in Carolina owed much to the employment of irregulars under and alongside his regular soldiers. Moreover, accounts of the war emphasise how the British found themselves entrapped 'by the all too often formidable combination of the regular Continental Army screened and supported by militias'.[36]

---

36 Jeffrey Record, 'External Assistance: Enabler of Insurgent Success', *Parameters* XXXVI:3 (2006), 36–49.

In highlighting this *positional* understanding of the differences between proxy and auxiliaries the chapter moves the debate beyond its current treatment of the issues according to which auxiliaries are essentially 'distinct from proxies, which are defined as receiving merely indirect support'.[37] In doing so, it addresses the under-studied character of the problem and places it into the broader issue of 'war amongst the people'.[38]

The chapter now turns to placing its theoretical observations into an empirical setting, namely Afghanistan and its long history of people's wars. As one of the paradigmatic contemporary cases of 'war amongst the people', Afghanistan witnessed both a conventionally soldier-heavy, regular military strategy[39] and a light footprint one, to which irregulars were key. Recent research has argued that the failure to address the Afghan strategic challenges post 9/11 was due in part to the reliance on security partnership with local allies whose unreliability became a strategic liability.[40] In drawing the proxy-auxiliary difference, the chapter points to the significance of this discussion as addressing a key determinant of success in contemporary war by offering an explanatory variant that underlies the importance of the distinction to successful warfighting.

The Proxy and Auxiliary Models in Practice: Afghanistan's 'Wars Amongst the People'

As Ayub and Kouvo argued, despite being richly documented, Afghanistan's wars are yet to be very well understood.[41] This can be observed in an oft-encountered push for explanations based on mere historical analogies.[42] Spanning across more than two centuries, Afghanistan's relationship with political violence has challenged historians, anthropologists, conflict researchers, political scientists and policy-makers alike.[43] In many ways its complex war history seems to

---

37 Scheipers, 'Irregular Auxiliaries after 1945', 16.
38 Scheipers, 'Irregular Auxiliaries after 1945', 14.
39 Artemy M. Kalinovsky, *A Long Goodbye: The Soviet Withdrawal from Afghanistan* (Cambridge: Harvard University Press, 2011).
40 Stephen Biddle, Julia Macdonald and Ryan Baker, 'Small Footprint, Small Payoff: The Military Effectiveness of Security Force Assistance', *Journal of Strategic Studies* 41:1/2 (2018), 89–142.
41 Fatima Ayub and Sari Kouvo, 'Righting the course? Humanitarian intervention, the war on terror and the future of Afghanistan', *International Affairs* 84:4 (2008), 641–57.
42 Paul D. Miller, 'Graveyard of Analogies: The Use and Abuse of History for the War in Afghanistan', *Journal of Strategic Studies* 39:3 (2016), 446–76.
43 Thomas J. Barfield, *Afghanistan: A Cultural and Political History* (Princeton: Princeton University Press, 2010); Robert C. Crews, *Afghan Modern: The History of a Global Nation* (Cambridge: Cambridge University Press, 2015).

have contradicted history's ever-present sense of chronology, with its commitment to linearity and belief in progress. Critically, Afghanistan postulates historical repetition not as error, but as specificity. Indeed, it is now a common practice to claim that the history of the Afghan wars exerted a certain magnetism for major power intervention. As Hilali remarked, 'since the 19th century … Afghanistan has continued to suffer from superpower politics, external pressure, and chronic instability'.[44] It is even more common to expand on the country's ability to reject it.[45] While Gibbs presented Afghanistan as one of the few countries in Central Asia never to be subject to direct colonial rule,[46] Yousaf and Adkin claimed that, with the exception of a shared religion, it was only foreign invaders that united the Afghans.[47]

What is less common is an appreciation of Afghanistan's multitude of inter-related conflicts – international, national and sub-national – in a way that demonstrates that political violence, expressed either directly or indirectly, rarely takes place in 'isolated pairs, but rather in a networked system'.[48] A key feature of this problem is the treatment of the uses and roles of the irregulars as proxy and auxiliaries in Afghan wars in a literature that otherwise has produced a veritable exegesis of its subject. It has so far been embedded in the broad narratives of strategic struggles: expansionist, geopolitical, ideological and religious. This marginalised the key differences between proxies and auxiliaries, as well as the conditions allowing for the two strategic models to evolve: the extensive Afghan tribal factionalism[49] and the structuring of social practices into kinship-based, patron-client relationships.

Taken together, these have placed the Afghans at the centre of a stream of wars waged *against, with, for, amongst* and *alongside* them. The Afghan people accommodated, waged and distributed violence among their patrons and between themselves, outside and alongside their customary

---

44   A.Z. Hilali, 'The Soviet Penetration into Afghanistan and the Marxist Coup', *The Journal of Slavic Military Studies* 18:4 (2005), 674.

45   Seth Jones, *In the Graveyard of Empires: America's War in Afghanistan* (New York: W.W. Norton & Co., 2010); David Isby, *Afghanistan: Graveyard of Empires: A New History of the Borderland* (New York: Pegasus, 2011).

46   David Gibbs, 'The Peasant as Counter-Revolutionary: The Rural Origins of the Afghan Insurgency', *Studies in Comparative International Development* 21:1 (1986), 47.

47   Mohammed Yousaf and Mark Adkin, *Afghanistan: The Bear Trap: The Defeat of a Superpower* (New York: Casemate, 2001), 128.

48   Sarah E. Croco and Tze Kwang Teo, 'Assessing the Dyadic Approach to Interstate Conflict Processes: A.k.a. "Dangerous" Dyad-Years', *Conflict Management and Peace Science* 22:5 (2005), 5–18.

49   Hilali, 'The Soviet Penetration into Afghanistan and the Marxist Coup', 680; see also Thomas H. Johnson and M. Chris Mason, 'No Sign until the Burst of Fire. Understanding the Pakistan-Afghanistan Frontier', *International Security* 32:4 (2008), 41–77.

practice of *badal* (vendetta)[50] or *tarburwali* (cousin rivalry).[51] Yet most analyses paint the Afghans as insurgents, with the existing appreciation of the Afghan irregular effort generically linked to the country's Cold War struggle, which has now become synonymous with the country's origins of proxy wars. That Afghanistan, 'more than any other location, was the high point of the Cold War',[52] is common knowledge, with the mujahedeen's fight against the Soviets central to any study of Cold War historiography. Galster, prefacing the National Security Archive's online volume of declassified documents, *Afghanistan: Lessons from the Last War*, called it the 'battleground for the bloodiest superpower proxy war of the 1980s'.[53] Similarly, Blum argued that 'Afghanistan was a cold-warrior's dream: The CIA and the Pentagon, finally, had one of their proxy armies in direct confrontation with the forces of the Evil Empire'.[54]

However, by the time the Soviet Union invaded Afghanistan, precipitating another West-East proxy war, the Afghan warriors found themselves in a familiar situation, albeit a century apart. The nineteenth century saw Afghanistan develop as a buffer state between the British and Russian empires. The second half of the chapter begins by presenting the proxy strategic model starting with the Soviet intervention of 1979. Against this background the auxiliary model will be compared and contrasted, observing the theoretical differences explained previously. Understanding the contemporary value of the two strategic models requires a longer historical perspective because it allows the development of an understanding of proxies and auxiliaries that is not context dependent, but instead shows their strategic utility through time.

Rubin's exceptional study on the transformation of Afghanistan over the last few decades argued that the country's encounters with the phenomenon of proxy war – and thus with the first strategic model presented here – began with the Cold War.[55] The key event is the Soviet intervention in Afghanistan in 1979. Afghanistan's geographical positioning vis-à-vis the Soviet Union had always produced a special

50 Gibbs, 'The Peasant as Counter-Revolutionary', 43.

51 Matthew Fielden and Jonathan Goodhand, 'Beyond the Taliban? The Afghan Conflict and United Nations Peacemaking', *Conflict, Security and Development* 1:3 (2001), 5–32.

52 Mahmood Mamdani, *Good Muslim, Bad Muslim: America, the Cold War, and the Roots of Terror* (New York: Pantheon Books, 2005), 120.

53 Steve Galster, *Afghanistan: The Making of U.S. Policy, 1973–1990* (Washington DC: George Washington University, 2001), available at http://nsarchive.gwu.edu/NSAEBB/NSAEBB57/essay.html, accessed 17 October 2018.

54 William Blum, *Killing Hope: U.S. Military and CIA Interventions since World War II* (London: Zed Books, 2014), 345.

55 Barnett R. Rubin, *Afghanistan from the Cold War through the War on Terror* (Oxford: Oxford University Press, 2015), 25.

relationship[56] and the threat of losing the country to the Americans altered the Soviet view of the strategic context, pushing for direct military intervention.[57] After all, the Chinese absorbed Tibet in 1951 and the US ran a path-breaking proxy war in Guatemala in 1954. Having an assertive power policy concerning one's own backyard – or front yard – was an ordering principle of the Cold War. For the US, on the other hand, Afghanistan, by itself, was of little importance.[58] However, the loss of Iran as their 'policeman' in the Middle East, following the overthrow of the Pahlavi dynasty in 1979, and the prospects of the Soviets becoming entangled in a Vietnam war of their own, informed the decision to be 'more sympathetic to those Afghans who were determined to preserve their country's independence'.[59] These were the mujahedeen, holy warriors aggregated in small battalions or *jabhas*[60] that would become, next to the contras in Nicaragua, the most recognisable proxy actors for the US. The proxy model began with President Jimmy Carter's modest efforts and was overridden by President Ronald Reagan's outspoken and overt support of the mujahedeen. The outside support brought Afghanistan's domestic conflict into the 'geopolitical logic of the Cold War'[61] and ensured that the process of building the mujahedeen army would be extensively traced.[62]

First, the weaving of the Afghan war throughout the 1980s into the broader geopolitical balance speaks of the politico-strategic role of the proxies, as presented previously. Indeed, for the US, supporting the

---

56 Blum, *Killing Hope*, 339.

57 The debate on the rationale behind the Soviet intervention is extensive. See Odd Arne Westad, 'Prelude to Invasion: The Soviet Union and the Afghan Communists, 1978–1979', *The International History Review* 16:1 (1994), 49–69; David N. Gibbs, 'Reassessing Soviet Motives for Invading Afghanistan: A Declassified History', *Critical Asian Studies* 38:2 (2006), 239–63; Artemy Kalinovsky, 'Decision-Making and the Soviet War in Afghanistan: From Intervention to Withdrawal', *Journal of Cold War Studies* 11:4 (2009), 46–73; Rodric Braithwaite, *Afgantsy: the Russians in Afghanistan 1979–89* (London: Profile Books, 2011); James D.J. Brown, 'Oil Fueled? The Soviet Invasion of Afghanistan', *Post-Soviet Affairs* 29:1 (2013), 56–94.

58 Galster, *Afghanistan: The Making of U.S. Policy*.

59 Zbigniew Brzezinski, *Power and Principle: Memoirs of the National Security Adviser, 1977–1981* (New York: Farrar, Straus, Giroux, 1983), 427.

60 A.Z. Hilali, 'Afghanistan: The Decline of Soviet Military Strategy and Political Status', *The Journal of Slavic Military Studies* 12:1 (1999), 104.

61 Fielden and Goodhand, 'Beyond the Taliban? The Afghan Conflict and United Nations Peacemaking', 8.

62 Charles G. Cogan, 'Partners in Time: The CIA and Afghanistan since 1979', *World Policy Journal* 10:2 (1993), 73–82; Alan J. Kuperman, 'Stinger Missile and U.S. Intervention in Afghanistan', *Political Science Quarterly* 114:2 (1999), 219–63; Steve Coll, *Ghost Wars: The Secret History of the CIA, Afghanistan, and Bin Laden, from the Soviet Invasion to September 10, 2001* (London: Penguin, 2005).

mujahedeen was part of their strategy of managing systemic relations with the USSR, which culminated in National Security Directive 166, whose ultimate goal was 'to push the Soviets out'.[63] Second, US support consisted of military and financial assistance to the mujahedeen as proxies. In the context of the proposed theorisation, the fighters fought for the expulsion of the Soviets from Afghanistan and did so *for* the US as well. This was not without problems. For one thing, using Pakistan as a conduit for transport and allocation of support diverted the strategic effort. Problems mounted as both sides searched for proxies. This was the case with the Hazara and the Afridi tribes, who, after being enlisted by the Soviets to stop the mujahedeen near the Pakistan border, turned against their patrons, 'trapping the Soviets in a crossfire with the resistance'.[64] Of relevance here is also the fact that the veiled expression of 'defeating the Soviet infidels' did little to help, for the resistance was undermined from inside by lack of unity, factionalism and splintering, as well as by repeated shifts in their support for each other.[65] The fiercely independent nature of the mujahedeen proved its strategic hubris and derailed both international and regional proxy wars.

Nevertheless, what this highlights in relation to the difference between auxiliaries and proxies is that the initial fighting dynamic is changed by the irregular force that assumes a grant of authority from a third party, whether it be the US, USSR or Pakistan. As discussed above, notwithstanding the negative course the model can assume, what is significant here is the extension of the party interactions to the inclusion of the so-called Beneficiary. In the context of Afghanistan in the 1980s, the political appeal overrode concerns over potentially negative outcomes to such an extent that sponsorship of militias and rebels was extensive and proxy war networks were established not just by the US, but also by the Soviets, the Afghan communist government, the UK, Iran, Saudi Arabia, Egypt and China, as well as by militias and local tribes.[66]

The Chinese, for example, clashing with the Soviets on ideological and territorial grounds, provided the mujahedeen with Chinese manufactured AK-47s.[67] Interestingly, strategic isolation in the international system and fears of Soviet encirclement informed China's involvement, despite

63 Robert M. Gates, *From the Shadows: The Ultimate Insider's Story of Five Presidents and How They Won the Cold War* (New York: Simon & Schuster, 1995), 319–21.
64 See Craig M. Karp, 'The War in Afghanistan', *Foreign Affairs* 64:5 (1986).
65 William Maley, 'The Future of Islamic Afghanistan', *Security Dialogue* 24:4 (1993), 386.
66 Lester W. Grau, 'The Soviet–Afghan War: A Superpower Mired in the Mountains', *The Journal of Slavic Military Studies* 17:1 (2004), 149.
67 Brian Glyn Williams, 'Afghanistan after the Soviets: From Jihad to Tribalism', *Small Wars and Insurgencies*, 25:5/6 (2014), 924–56.

remaining largely unacknowledged.[68] The UK established a strong relationship with Ahmed Shah Massoud, the 'Lion of Panjshir'.[69] The Afghan government financed the members of the Hazara tribe who, a century before, had been bludgeoned, first by the British, who punished them for refusing to sell fodder by burning their fields,[70] and, later, by Abdur Rahman when trying to forge the Afghan state. This Islamic Shiite minority guarded the Hindu Kush Mountains and 'went even further and actively fought for the Communist government against their hereditary enemies, the Pashtun mujahedeen'.[71] However, their strategic aim was political survival and so, to maximise its success, it turned to the new regime in Iran. As Iran stepped in, waging yet another proxy war in this already complex network of conflicts, the homogeneity of the Hazaras proved essential. By supporting Hazara religious leaders, Iran assisted in constructing an effective political administration,[72] once again showing the politico-strategic utility of proxies. However, as the tribe reconfigured along ethnic roots at the expense of its religious outlook in the wake of the Soviet withdrawal, 'Iran decided to accept the fact that its Hazara proxies would not be able to establish an Iranian-style regime in the heart of Afghanistan'.[73]

As anticipated by the US, the Soviets had envisioned a short intervention without ever imagining they would be 'involved in the middle of a civil war on extremely rugged terrain where the Soviets … would carry the bulk of the combat burden'.[74] Ten years later, as they withdrew, Afghanistan collapsed into a civil war where 'the mujahedeen, along with the remnants of the army turned into feuding warlords and ethnic militias'.[75] The conflict soon became even more multifaceted, pulling and pushing local, national and regional actors in the violent web of small and quickly shifting proxy wars. As Fielden and Goodhand remarked, the Afghan conflict could be characterised as 'part regional proxy war and part civil war' for it has 'shifted from a bipolar war to a multipolar regional

68 Jonathan Z. Ludwig, 'Sixty Years of Sino-Afghan Relations', *Cambridge Review of International Affairs* 26:2 (2013), 392–410.
69 Geraint Hughes, 'The Soviet–Afghan War, 1978–1989: An Overview', *Defence Studies* 8:3 (2008), 328; Panagiotis Dimitrakis, 'The Soviet Invasion of Afghanistan: International Reactions, Military Intelligence and British Diplomacy', *Middle Eastern Studies* 48:4 (2012), 511–36.
70 Vartan Gregorian, *The Emergence of Modern Afghanistan: Politics of Reform and Modernization, 1880–1946* (Stanford: Stanford University Press, 1969), 109.
71 Williams, 'Afghanistan after the Soviets: From Jihad to Tribalism', 934.
72 Karp, 'The War in Afghanistan', 1030.
73 Williams, 'Afghanistan after the Soviets: From Jihad to Tribalism', 943.
74 Grau, 'The Soviet–Afghan War: A Superpower Mired in the Mountains', 129–51.
75 Rubin, *Afghanistan from the Cold War through the War on Terror*, 25.

proxy war, involving neighbouring powers, China, Iran, Pakistan and the Central Asian Republics'.[76] Rubin expressed a similar view that, after the collapse of the Soviet Union, multiple funding channels emerged, some of which involved non-state actors.[77] The conflict moved from an international level with the end of the bipolar system, into a regional one to an extent that surpassed the regional involvement of the 1980s.[78] The proxy relations swiftly shifted in terms of strategic content, pushing the 'war amongst the Afghan people' into a veritable web of 'wars amongst the people' of South and Central Asia in a way that demonstrates how the proxy model changes the dynamics of 'war amongst the people' in general. While some groups, such as the one led by Gulbuddin Hekmatyar, continued their relationship with Pakistan and Saudi Arabia, those led by Burhahuddin Rabbani and Ahmed Shah Massoud welcomed Russia, Iran and India and their own policies of covert aid, as was theorised above.[79]

In the context of the discussion of the proxy model as part of the wider issue of 'war amongst the people', this had tremendous implications. In line with what Smith argued, the increasing plethora of non-state actors took a distinctively active and political role. Earlier, the chapter noted Iran's sponsorship of the Hazaras as a proxy. Before the end of the Soviet intervention, they had been a conduit for a proxy war against Saudi Arabia and Riyadh's support for Wahhabism in Afghanistan.[80] However, in search of political survival and representation, the Hazaras reconfigured themselves politically and socially, slowly rescinding even the indirect cooperation with Iran. The agency of local actors and its pursuit through strategic interaction is evident here, as in the case of the mujahedeen, for the Hazaras transformed the jihad of Hazarajat into the plight of an ethnic-based movement.[81] This also qualifies the implications of working *for* a Beneficiary as a strategic model subject to volatility and rapid shifts. As such, by the time the US intervened in Afghanistan in 2001 and itself began providing support for anti-Taliban militias and warlords – albeit to a different end – a complex combination of states and non-state actors already had a two-decade-long history of using proxy forces, either Afghan mujahedeen or the Taliban. Having drawn a picture of the evolution and implications of the proxy model, the chapter now

---

76 Fielden and Goodhand, 'Beyond the Taliban? The Afghan Conflict and United Nations Peacemaking', 6.
77 Rubin, *Afghanistan from the Cold War through the War on Terror*, 30.
78 Anwar-ul-Haq Ahady, 'The Decline of the Pashtuns in Afghanistan', *Asian Survey* 35:7 (1995), 621–34.
79 Rubin, 'Women and Pipelines. Afghanistan's Proxy Wars', 286.
80 Maley, 'The Future of Islamic Afghanistan', 390.
81 Williams, 'Afghanistan after the Soviets: From Jihad to Tribalism', 944.

turns to detailing the specificities of the auxiliary model, which became a feature of the 2001 intervention.

Through its intervention in Afghanistan, the US sought to create an inhospitable base for extremism, which, as an aim, was different to the country's historical experience of foreign interventions. The aim paled in comparison and effort to that of the Soviet Union and the fear of burying itself in the 'graveyard of empires'[82] impacted significantly on the shape of US strategy: a combination of airpower and a light footprint *with* the cooperation of local forces. Having dislodged the Taliban from the official seats of power by 2002, the war effort concentrated henceforth on stabilisation and defeat of an ebbing and flowing insurgency. During these phases, the power of local entrepreneurs of violence was harnessed in accordance with the precepts of American counterinsurgency as *auxiliaries*. In this case, the Northern Alliance came to be the auxiliary prototype. However, it is important to note that the chapter focuses on the irregular forces that are co-opted to work with the regular one in what can be called informal tactical alliances and not with grass-roots forces that end up subordinated and embedded in local, regional or national structures of authority. This would be the case of the local defence forces who, as Strandquist showed, despite being effectively tribal militias aimed at fighting the Taliban, remained subordinated to the central government as part of the Community Defence Initiative.[83] It also does not include official forces, such as the Afghan National Auxiliary Police (ANAP), Afghan Local Police Program or the development and employment of the Afghan National Army (ANA) as a US auxiliary force.[84] There is indeed considerable overlap between the roles these forces play and those of the irregular auxiliaries within the counterinsurgency spectrum of operations, but they are not the focus here.

The use of irregular auxiliaries became a key provision of Western doctrine in the aftermath of the US intervention in Afghanistan. The 2004 *Field Manual Interim 3-07.22, Counterinsurgency Operations* highlighted the imperative to expand and employ strong and able native forces.[85] Similarly, the 2006 *US National Security Strategy* emphasised

82 Jones, *In the Graveyard of Empires: America's War in Afghanistan*.

83 Jon Strandquist, 'Local Defence Forces and Counterinsurgency in Afghanistan: Learning from the CIA's Village Defence Program in South Vietnam', *Small Wars and Insurgencies* 26:1 (2013), 90–113.

84 Antonio Giustozzi, 'Auxiliary Force or National Army? Afghanistan's 'ANA' and the Counter-Insurgency Effort, 2002–2006', *Small Wars and Insurgencies* 18:1 (2007), 45–67.

85 US Army, *Field Manual Interim 3-07.22, Counterinsurgency Operations* (Washington DC: US Army, 2004), available at https://fas.org/irp/doddir/army/fmi3-07-22.pdf, accessed 17 October 2018.

the importance of working with allies in order to develop capable indigenous security forces able to fight terrorist and insurgent threats.[86] For the UK, General Sir Michael Jackson acknowledged that the use of local indigenous forces, either inherited or built up *ab initio,* had been of increasing importance.[87] The relationship with such auxiliaries has also been scrutinised carefully, the role of irregulars often described as following an informal security and military contracting pattern run from the shadows by certain US government institutions, chiefly the CIA and the DoD:

> For more than a decade, wads of American dollars packed into suitcases, backpacks and, on occasion, plastic shopping bags have been dropped off every month or so at the offices of Afghanistan's president—courtesy of the Central Intelligence Agency.[88]

Critically, however, while detailed, criticised or revered, a common problem was the lack of conceptual clarity as to both the functions the local forces carried out and their capacity to undertake such tasks. More precisely, their role was presented in both research and policy in an interchangeable fashion as either proxy or auxiliary.[89] This was the case for the Northern Alliance and other irregulars. The baseline argument saw them operating once US airpower degraded the theatres of war and with US Special Forces as a screen against enemy attack.[90] In both the battle of Tora Bora and Operation Anaconda in the Shah-e-Knot valley, local forces were used. They were part of the light footprint strategy and were employed in various roles, such as launching attacks on enemy targets or, during Anaconda, to act as shock troops whose effort was aimed at uprooting al-Qaeda fighters from their bases. As detailed analyses of these key points in the war showed, the local forces' tactical skills were critical.[91] As such, they worked with and alongside and

---

86   White House, *National Security Strategy of the United States of America* (Washington DC: White House, 2006), available at https://www.state.gov/documents/organization/64884. pdf, accessed 17 October 2018.

87   General Sir Michael Jackson, 'British Counter-Insurgency', *Journal of Strategic Studies* 32:3 (2009), 347–51.

88   Matthew Rosenberg, 'With Bags of Cash, C.I.A. Seeks Influence in Afghanistan', *The New York Times*, 28 April 2013, available at http://www.nytimes.com/2013/04/29/world/asia/cia-delivers-cash-to-afghan-leaders-office.html, accessed 16 October 2018.

89   Richard B. Andres, Craig Wills and Thomas E. Griffith Jr, 'Winning with Allies: The Strategic Value of the Afghan Model', *International Security* 30:3 (2005), 124–60.

90   Andres, Wills, and Griffith Jr, 'Winning with Allies: The Strategic Value of the Afghan Model', 126.

91   Andres, Wills and Griffith Jr., 'Winning with Allies: The Strategic Value of the Afghan Model', 153.

assumed the military-tactical value of an auxiliary as described above. They became part of the official US strategy as a tactical complement and thus conserved the numbers of parties involved in fighting in a different way to the proxy strategic model. Seen through the lens of 'war amongst the people' the auxiliary model points to the emergence of multiple pathways through which the 'people' react *to* and *in* such war contexts. The general understanding of the 'people' as either insurgent or support base is changed and both the auxiliary and proxy models demonstrate how insurgencies shift as the fighting assumes new courses.

The difference between auxiliaries and proxies becomes clear during the years following the American intervention in 2001. While the invading regular forces developed close cooperation with auxiliary forces, such as the Northern Alliance, the proxy model took a distinctive regional turn. From the very moment the Taliban emerged victorious from the civil war in 1996, Pakistan assumed charge of the strategic bargaining through proxy wars to such an extent that violent dynamics in Afghanistan came to be all about Pakistan as well.[92] As Jones put it, 'the link to Pakistan was not a surprise, though the reality of outside support was much darker'.[93] Pakistan's wielding of proxy wars was both inward and outward looking and saw an important shift from collusion with the heroes of the Cold War to the Taliban.[94] Internally, Pakistan was driven by the imperatives of preserving state boundaries in the face of secessionist threats. Preoccupation with domestic stability reacted to and made recourse of sub-state ethnic groupings and the Taliban came to be the response to the Pashtun problem, as well as an instrument of Pakistani policy.[95] It was the very lens of the proxy strategic model that became the key to the current regional dialogue aimed at ending the Afghan-Taliban conflict. As a response to Pakistan's interference in the Afghan conflict, the Afghan government had, for a long time, supported Pakistani rebel groups, especially the Tehrik-e-Taliban. Using the proxy as a bargaining chip in combination with a mix of political and diplomatic moves, President Ghani slowly pushed for a rapprochement with Pakistan aimed at bringing the Taliban to the negotiating table, which currently continues to develop at a slow pace. By understanding the critical value

---

92 Greg Mills and Ewen Mclay, 'A Path to Peace in Afghanistan: Revitalizing Linkage in Development, Diplomacy and Security', *Orbis* 55:4 (2011), 605.

93 Seth G. Jones, *Waging Insurgent Warfare: Lessons from the Vietcong to the Islamic State* (Oxford: Oxford University Press, 2017), 3.

94 Ayub and Kouvo, 'Righting the course? Humanitarian intervention, the war on terror and the future of Afghanistan', 643; see also Carlota Gall, *The Wrong Enemy. America in Afghanistan, 2001–2014* (New York: Mariner Books Houghton Mifflin Harcourt, 2015).

95 Rubin, 'Transnational Justice and Human Rights in Afghanistan', 69.

of the extension of its 'war amongst the people' into wars in which the people participate in various ways, Afghanistan's president has sought to shift decades of 'wars amongst the Afghan people' towards potential peace.

Conclusion

This chapter has sought to use the case of Afghanistan's 'wars amongst the people' in order to determine two models of strategic interaction that have shaped both the course and outcomes of these conflicts. In drawing a distinction between the proxy and auxiliary models, the chapter emphasises the complexity of contemporary 'wars amongst the people', which, as Angstrom and Honig observed, are never conducted in a vacuum, but encompass a complicated set of actors who reproduce their interests in multiple ways.[96] To this end, strategic interaction became a lens for locating the differences between the auxiliary and proxy models. The starting assumption was that, in the case of Afghanistan, most of the time auxiliaries had been conceptualised as proxies and vice-versa under the pressures of a tightly defined political context.[97] Yet the problem was far more pressing because it ignored the degree to which the two sets of dynamics changed the character of 'war amongst the people' as strategic environments for which strategic solutions are sought and implemented. The chapter shows how two models of regular–irregular interaction fragmented and segmented Afghanistan and, more importantly, for how long. One of the most recent expressions of this fragmentation process came in April 2017, when former Afghan mujahedeen Gulbuddin Hekmatyar issued a call for peace in Afghanistan, inviting Afghans to 'join the peace caravan and stop the pointless, meaningless and unholy war'.[98] The archetype of the proxy warrior – who waged holy war against the Soviets, who tried to rule Afghanistan through and with Pakistani support and who finally rebelled against the Taliban – was now pursuing a radically different strategy focused on ending wars. Such shifts demonstrate the complex nature of 'war amongst the people', which are made possible by the very extension of these wars though proxy and auxiliary dynamics.

---

96 Jan Angstrom and Jan Willem Honig, 'Regaining Strategy: Small Powers, Strategic Culture, and Escalation in Afghanistan', *Journal of Strategic Studies* 35:5 (2012), 663–87.

97 Betz and Anthony Cormack, 'Iraq, Afghanistan and British Strategy', 322.

98 'Notorious Afghan Warlord Calls for Peace in First Public Speech', *Reuters*, 29 April 2017, available at https://www.reuters.com/article/us-afghanistan-hekmatyar/notorious-afghan-warlord-calls-for-peace-in-first-public-speech-idUSKBN17V08M, accessed 17 October 2018.

As they have been a feature of such wars for decades now, both military and scholarly thinking should assess their strategic value and, more importantly, their consequences, with greater precision and acuity.

References

Ahady, A-ul-Haq, 'The Decline of the Pashtuns in Afghanistan', *Asian Survey* 35:7 (1995).

Andres, R., Wills, C. and Griffith, T., 'Winning with Allies: The Strategic Value of the Afghan Model', *International Security* 30:3 (2005).

Angstrom, J., 'Inviting the Leviathan: external forces, war, and state-building in Afghanistan', *Small Wars and Insurgencies* 19:3 (2008).

Angstrom, J. and Honig, J., 'Regaining Strategy: Small Powers, Strategic Culture, and Escalation in Afghanistan', *Journal of Strategic Studies* 35:5 (2012).

Ayub, F. and Kouvo, S., 'Righting the course? Humanitarian intervention, the war on terror and the future of Afghanistan', *International Affairs* 84:4 (2008).

Barfield, T., *Afghanistan: A Cultural and Political History* (Princeton: Princeton University Press, 2010).

Betz, D. and Cormack, A., 'Iraq, Afghanistan and British Strategy', *Orbis* 53:2 (Spring 2009).

Betts, R.K., 'Is Strategy an Illusion?', *International Security* 25:2 (2000).

Biddle, S., 'Afghanistan's Legacy: Emerging Lessons of an Ongoing War', *The Washington Quarterly* 37:2 (2014).

Biddle, S., Macdonald, J. and Baker, R., 'Small Footprint, Small Payoff: The Military Effectiveness of Security Force Assistance', *Journal of Strategic Studies* 1:1/2 (2018).

Bird, T. and Marshall, A., *Afghanistan: How the West Lost Its Way* (Yale: Yale University Press, 2011).

Blum, W. *Killing Hope: U.S. Military and CIA Interventions since World War II* (London: Zed Books, 2014).

Braithwaite, R., *Afgantsy: the Russians in Afghanistan 1979–89* (London: Profile Books, 2011).

Brown, J., 'Oil Fueled? The Soviet Invasion of Afghanistan', *Post-Soviet Affairs* 29:1 (2013).

Brown, S., 'Purposes and Pitfalls of War by Proxy: A Systemic Analysis', *Small Wars and Insurgencies* 27:2 (2016).

Brzezinski, Z. *Power and Principle: Memoirs of the National Security Adviser, 1977–1981* (New York: Farrar, Straus & Giroux, 1983).

Chaudhuri, R. and Farrell, T., 'Campaign disconnect: operational progress and strategic obstacles in Afghanistan, 2009–2011', *International Affairs* 87:2 (2011).

Cogan, C., 'Partners in Time: The CIA and Afghanistan since 1979', *World Policy Journal* 10:2 (1993).

Coll, S., *Ghost Wars: The Secret History of the CIA, Afghanistan, and Bin Laden, from the Soviet Invasion to September 10, 2001* (London: Penguin, 2005).

Crews, R., *Afghan Modern: The History of a Global Nation* (Cambridge: Cambridge University Press, 2015).

Croco, S. and Teo, T., 'Assessing the Dyadic Approach to Interstate Conflict Processes: A.k.a. "Dangerous" Dyad-Years', *Conflict Management and Peace Science* 22:5 (2005).

Cunningham, D., Gleditsch, K. and Salehyan, I., 'It Takes Two: A Dyadic Analysis of Civil War Duration and Outcome', *Journal of Conflict Resolution* 53:4 (2009).

Dimitrakis, P., 'The Soviet Invasion of Afghanistan: International Reactions, Military Intelligence and British Diplomacy', *Middle Eastern Studies* 48:4 (2012).

Dodge, T., 'Intervention and dreams of exogenous statebuilding: the application of Liberal Peacebuilding in Afghanistan and Iraq', *Review of International Studies* 39 (2013).

Egnell, R., 'Lessons from Helmand, Afghanistan: what now for British counterinsurgency?', *International Affairs* 87:2 (2011).

Farrell, T. and Giustozzi, A., 'The Taliban at war: inside the Helmand insurgency, 2004–2012', *International Affairs* 89:4 (2013).

Farrell, T. and Semple, M., 'Making Peace with the Taliban', *Survival* 57:6 (2015).

Fielden, M. and Goodhand, J., 'Beyond the Taliban? The Afghan Conflict and United Nations Peacemaking', *Conflict, Security and Development* 1:3 (2001).

Freedman, L. *Strategy: A History* (Oxford: Oxford University Press, 2013).

Gall, C., *The Wrong Enemy. America in Afghanistan, 2001–2014* (New York: Mariner Books Houghton Mifflin Harcourt, 2015).

Galster, S., *Afghanistan: The Making of U.S. Policy, 1973–1990* (Washington DC: George Washington University, 2001), available at http://nsarchive.gwu.edu/NSAEBB/NSAEBB57/essay.html, accessed 17 October 2018.

Gates, R.M. *From the Shadows: The Ultimate Insider's Story of Five Presidents and How They Won the Cold War* (New York: Simon & Schuster, 1995).

Gibbs, D., 'The Peasant as Counter-Revolutionary: The Rural Origins of the Afghan Insurgency', *Studies in Comparative International Development* 21:1 (1986).

Gibbs, D., 'Reassessing Soviet Motives for Invading Afghanistan: A Declassified History', *Critical Asian Studies* 38:2 (2006).

Giustozzi, A., 'Auxiliary Force or National Army? Afghanistan's 'ANA' and the Counter-Insurgency Effort, 2002–2006', *Small Wars and Insurgencies* 18:1 (2007).

Grau, L., 'The Soviet–Afghan War: A Superpower Mired in the Mountains', *The Journal of Slavic Military Studies* 17:1 (2004).

Gregorian, V. *The Emergence of Modern Afghanistan: Politics of Reform and Modernization, 1880–1946* (Stanford: Stanford University Press, 1969).

Griffin, S., 'Iraq, Afghanistan and the future of British military doctrine: from counterinsurgency to Stabilization', *International Affairs* 87:2 (2011).

Heuser, B. (ed.), *Small Wars and Insurgencies in Theory and Practice, 1500–1850* (London and New York: Routledge, 2017).

Hilali, A., 'Afghanistan: The Decline of Soviet Military Strategy and Political Status', *The Journal of Slavic Military Studies* 12:1 (1999).

Hilali, A., 'The Soviet Penetration into Afghanistan and the Marxist Coup', *The Journal of Slavic Military Studies* 18:4 (2005).

Hughes, G., 'The Soviet–Afghan War, 1978–1989: An Overview', *Defence Studies* 8:3 (2008).

Hughes, G., *My Enemy's Enemy: Proxy Warfare in International Politics* (Eastbourne: Sussex Academic Press, 2012).

Innes, M. (ed.), *Making Sense of Proxy Wars: States, Surrogates and the Use of Force* (Washington DC: Potomac Books, 2012).

Isby, D., *Afghanistan: Graveyard of Empires: A New History of the Borderland* (New York: Pegasus, 2011).

Jackson, M., 'British Counter-Insurgency', *Journal of Strategic Studies* 32:3 (2009).

Johnson, T. and Mason, M., 'No Sign until the Burst of Fire. Understanding the Pakistan-Afghanistan Frontier', *International Security* 32:4 (2008).

Jones, S., *In the Graveyard of Empires: America's War in Afghanistan* (New York: W.W. Norton & Co., 2010).

Jones, S., *Waging Insurgent Warfare: Lessons from the Vietcong to the Islamic State* (Oxford: Oxford University Press, 2017).

Kalinovsky, A., 'Decision-Making and the Soviet War in Afghanistan: From Intervention to Withdrawal', *Journal of Cold War Studies* 11:4 (2009).

Kalinovsky, A., *A Long Goodbye: The Soviet Withdrawal from Afghanistan* (Cambridge: Harvard University Press, 2011).

Karp, C., 'The War in Afghanistan', *Foreign Affairs* 64:5 (1986).

Koehler, K., Ohl, D. and Albrecht, H., 'From Disaffection to Desertion: How Network Facilitate Military Insubordination in Civil Conflict, *Comparative Politics* 48:4 (2016).

Krieg, A., 'Externalizing the Burden of War: the Obama Doctrine and US Foreign Policy in the Middle East', *International Affairs* 92:1 (2016).

Kuperman, A.J., 'Stinger Missile and U.S. Intervention in Afghanistan', *Political Science Quarterly* 114:2 (1999).

Lake, D.A. and Powell, R. (eds), *Strategic Choice and International Relations* (Princeton: Princeton University Press, 1993).

Ledwidge, F., *Losing Small Wars: British Military Failure in the 9/11 Wars* (Yale: Yale University Press, 2017).

Ludwig, J.Z. 'Sixty Years of Sino-Afghan Relations', *Cambridge Review of International Affairs* 26:2 (2013).

Maley, W., 'The Future of Islamic Afghanistan', *Security Dialogue* 24:4 (1993).

Mamdani, M., *Good Muslim, Bad Muslim: America, the Cold War, and the Roots of Terror* (New York: Pantheon Books, 2005).

Maoz, Z. and San-Akca, B., 'Rivalry and State Support of Non-State Armed Groups (NAGs), 1946–2001', *International Studies Quarterly* 56 (2012).

Marks, T. and Rich, P., 'Back to the Future: People's War in the 21st Century', *Small Wars and Insurgencies* 28:3 (2017).

Marsden, P., 'Afghanistan: the Reconstruction Process', *International Affairs* 79:1 (2003).

Marshall, A., 'From Civil War to Proxy War: Past History and Current Dilemmas', *Small Wars and Insurgencies* 27:2 (2016).

Miller, P.D., 'Graveyard of Analogies: The Use and Abuse of History for the War in Afghanistan', *Journal of Strategic Studies* 39:3 (2016).

Miller, P.D. 'On Strategy, Grand and Mundane', *Orbis* 60:2 (2016).

Mills, G. and Mclay, E., 'A Path to Peace in Afghanistan: Revitalizing Linkage in Development, Diplomacy and Security', *Orbis* 55:4 (2011).

Mumford, A., *Proxy Warfare* (London: Polity, 2013).

Rauta, V., 'Proxy Agents, Auxiliary Forces, and Sovereign Defection: Assessing the Outcomes of Using Non-State Actors in Civil Conflicts', *Southeast European and Black Sea Studies* 16:1 (2016).

Rauta, V. and Mumford, A. 'Proxy Wars and the Contemporary Security Environment' in R. Dover, H. Dylan and M.S. Goodman (eds), *The Palgrave Handbook of Security, Risk and Intelligence* (London: Palgrave Macmillan, 2017).

Record, J, 'External Assistance: Enabler of Insurgent Success', *Parameters* XXXVI:3 (Autumn 2006).

Romano, D., Calfano, B. and Phelps, R., 'Successful and Less Successful Interventions: Stabilizing Iraq and Afghanistan', *International Studies Perspectives* 16 (2015).

Rubin, B.R., 'Women and Pipelines. Afghanistan's Proxy Wars', *International Affairs* 73:2 (1997).

Rubin, B.R., 'Transnational Justice and Human Rights in Afghanistan', *International Affairs* 79:3 (2003).

Rubin, B.R., *Afghanistan from the Cold War through the War on Terror* (Oxford: Oxford University Press, 2015).

San-Akca, B., *States in Disguise: Causes of State Support for Rebel Groups* (Oxford: Oxford University Press, 2016).

Scheipers S., 'Counterinsurgency or Irregular Warfare? Historiography and the Study of Small Wars', *Small Wars and Insurgencies* 25:5/6 (2014).

Scheipers, S., *Unlawful Combatants: A Genealogy of the Irregular Fighter* (Oxford: Oxford University Press, 2015).

Scheipers, S., 'Irregular Auxiliaries after 1945', *The International History Review* 39:1 (2017).

Smith, R., *The Utility of Force: The Art of War in the Modern World* (London: Penguin, 2006).

Staniland, P., 'States, Insurgents, and Wartime Political Orders', *Perspectives on Politics* 10:2 (2012).

Staniland, P., 'Armed Groups and Militarized Elections', *International Studies Quarterly* 59 (2015).

Strachan, H. 'The Lost Meaning of Strategy', *Survival* 47:3 (2005).

Strandquist, J., 'Local Defence Forces and Counterinsurgency in Afghanistan: Learning from the CIA's Village Defence Program in South Vietnam', *Small Wars and Insurgencies* 26:1 (2013).

US Army, *Field Manual Interim 3-07.22, Counterinsurgency Operations* (Washington DC: US Army, 2004), available at https://fas.org/irp/doddir/army/fmi3-07-22.pdf, accessed 17 October 2018.

Westad, O.A., 'Prelude to Invasion: The Soviet Union and the Afghan Communists, 1978–1979', *The International History Review* 16:1 (1994).

White House, *National Security Strategy of the United States of America* (Washington DC: White House, 2006), available at https://www.state.gov/documents/organization/64884.pdf, accessed 17 October 2018.

Williams, B., 'Afghanistan after the Soviets: From Jihad to Tribalism', *Small Wars and Insurgencies* 25:5/6 (2014).

Yousaf, M. and Adkin, M. *Afghanistan: The Bear Trap: The Defeat of a Superpower* (New York: Casemate, 2001).

6

ENHANCING OPERATIONAL EFFECTIVENESS?

Pre-deployment Training for Tactical-level Rwandan Female Military Peacekeepers

Georgina Holmes

Introduction

UN Security Council Resolution 1325, its related resolutions and recent high-level reviews emphasise the unique ways women enhance peacekeeping activities by introducing gender sensitivity into tasks and during interactions with local women in the host country; assisting survivors of conflict-related sexual violence (CRSV); demobilising female ex-combatants and speaking with women from cultures that forbid their interaction with men.[1] Critics of the instrumentalist perspective rightly challenge the normative assumption that the operational value of women is dependent on their sex-difference. They also caution against counting women and essentialising female identities as peaceful and non-threatening and call for the transformation of male-dominated, militarised working cultures that subjugate female

---

1 See Gerard Jan De Groot, 'A few good women: Gender stereotypes, the military and peacekeeping', *International Peacekeeping* 8:2 (2001), 23–38; Louise Olsson and Torunn Tryggestad, *Women and International Peacekeeping* (London/New York: Routledge, 2001); Donna Bridges and Debbie Horsfall, 'Increasing Operational Effectiveness in UN Peacekeeping: Toward a Gender-Balanced Force', *Armed Forces and Society* 36:1 (2009), 120–30; Natasja Rupesinghe, John Karlsrud and Eli Stamnes, 'Women, Peace and Security and Female Peacekeeping Personnel' in Sara Davies and Jacqui True (eds), *The Oxford Handbook of Women, Peace and Security* (Oxford: Oxford University Press, 2019).

peacekeepers.[2] Yet, how tactical-level female military peacekeepers are trained, equipped and socialised by Troop Contributing Countries (TCCs) to work in mixed-gender contingents requires investigating, since TCCs are likely to operationalise the UN's gender mainstreaming norms differently during the pre-deployment training process.

In this chapter, French sociologist Pierre Bourdieu's theory of social practice is used to examine how UN gender mainstreaming norms (re) produced in Department of Peace-Keeping Operations (DPKO) and Department of Field Support (DFS) training materials are operationalised by the Rwandan Defence Force (RDF) during the pre-deployment training for tactical-level Rwandan military peacekeepers. It considers the consequences that this operationalisation has for determining how female peacekeepers are equipped, prior to deployment. Rwanda is chosen because the small African state has been deploying troops to difficult missions including the United Nations – African Union Hybrid Operation in Darfur (UNAMID) and the UN Mission in the Republic of South Sudan (UNMISS) since 2004. Over the past five years the TCC has contributed some 6,000 troops annually, of which 250 (4 per cent) are women.[3] Rwanda also has a UNSCR 1325 National Action Plan (2009), suggesting that the TCC is willing to integrate a gender perspective into its domestic and international security policies.[4] Rwanda's history of genocide and civil war in the 1990s has also contributed to the Government of Rwanda's commitment to improving and delivering on the UN's Protection of Civilian (PoC) mandate. Despite its own military

---

2   See Liora Sion, 'Peacekeeping and the Gender Regime: Dutch female peacekeepers in Bosnia and Kosovo', *Journal of Contemporary Ethnography* 37:5 (2008), 561–85; Gina Heathcote and Diane Otto, *Rethinking Peacekeeping, Gender Equality and Collective Security* (London: Palgrave Macmillan, 2014); Oliviera Simic, 'Increasing Women's Presence in Peacekeeping Operations: The Rationales and Realities of 'Gender Balance' in Gina Heathcote and Diane Otto (eds), *Rethinking Peacekeeping, Gender Equality and Collective Security* (London: Palgrave Macmillan, 2014); Kyle Beardsley and Sabrina Karim, 'Female peacekeepers and gender balancing: Token gestures or informed policymaking?', *International Interactions* 39:4 (2013), 461–88; Lesley J. Pruitt, *The Women in Blue Helmets: Gender, Policing and the UN's First All-Female Peacekeeping Unit* (Oakland: University of California Press, 2016); Sabrina Karim and Kyle Beardsley, *Equal Opportunity Peacekeeping* (Oxford: Oxford University Press, 2017).

3   Statistics provided by the RDF Gender Desk, 9 July 2015.

4   For a critique of the RDF's gender and security sector reform initiatives, see Annika Björkdahl and Johanna Mannergren Selimovic, 'Translating UNSCR 1325 from the global to the local: protection, representation and participation in the National Action Plans of Bosnia-Herzegovina and Rwanda', *Conflict, Security and Development* 14:4 (2015), 311–35; Georgina Holmes, 'Gendering the Rwanda Defence Force: A Critical Assessment', *Journal of Intervention and Statebuilding* 8:4 (2014), 321–33; Georgina Holmes, 'Gender and the Military in Post-Genocide Rwanda' in Elissa Bemporad and Joyce W. Warren (eds), *Women and Genocide: Survivors, Victims, Perpetrators* (Bloomington: Indiana University Press, 2018), 223–49.

intervention and unlawful economic activities in the east of Congo since the 2000s onwards, the political elite, led by President Paul Kagame and former Rwanda Patriotic Front (RPF) military leaders, has long been critical of the international community's failure to intervene to halt genocide. Within the UN, this failure to intervene in Rwanda was a key factor that led to the emergence of a new 'cosmopolitan' military doctrine in the late 1990s, following recognition of the need to place civilians, who were increasingly being targeted, at the centre of conflict prevention and conflict resolution, in line with the precepts of 'war amongst the people'. In cosmopolitan peacekeeping, forces are expected to act as 'security guarantors' that 'help defend' and 'restore' civil society, while supporting local legitimacy and 'pluralist democratic practices'.[5] Yet individual peacekeepers are also expected to support development projects, such as Quick Impact Projects (QIP), and build strong relations with the local population in order to help secure the mission's legitimacy in the host state. Female peacekeepers in particular are considered to help achieve these strategic objectives by engaging with women and vulnerable groups from local communities. Rwanda is also a key advocate of the Kigali Principles on the Protection of Civilians, which are a set of non-binding pledges drawn up in 2015. The principles call for the implementation of best practice in peacekeeping to improve the protection of local communities during protracted 'wars amongst the people'.

In this chapter, two interrelated research questions are posed: 1) How are gendered peacekeeping subject positions constructed through pre-deployment training programmes? 2) How do these constructions enable or constrain female military peacekeepers once they reach the theatre of operations? It is argued that a 'scripted' Rwandan female peacekeeper subject position is constructed during the pre-deployment programme, which is based on how RDF senior officers perceive women's value within a given mission. RDF institutional culture, norms and training practices introduce additional structural limitations into the UN's training curriculum, which prevent their tactical-level female military personnel from being prepared for the more challenging situations that they find themselves in when working in multi-dimensional peace operations.

The pre-deployment training space is first conceptualised as a site of norm contestation, wherein gendered peacekeeping subject positions are constructed. The chapter then examines how gender mainstreaming norms are translated in DPKO/DFS policy discourse and Core Pre-deployment

---

5   Lorraine Elliot, 'Cosmopolitan ethics and militaries as "good forces"' in Lorraine Elliot and Graeme Cheeseman (eds), *Forces for Good: Cosmopolitan Militaries in the Twenty-first Century* (Manchester: Manchester University Press, 2004), 25.

Training Materials (CPTMs). How these gender mainstreaming norms are operationalised during pre-deployment training for tactical-level Rwandan military personnel deployed to UNAMID in 2014–2015 is then analysed, along with Rwandan female military peacekeepers' perceptions of their training experiences. The chapter concludes by reflecting on how UN DPKO/DFS training materials and methods could be improved in order to better equip female peacekeepers, in light of the expectation that women support delivery of cosmopolitan peacekeeping objectives as a response to 'wars amongst the people'.[6]

Preparing Peacekeepers for Deployment

The shift from traditional peacekeeping missions towards multi-dimensional peacekeeping missions from the late 1990s has resulted in the articulation of two types of peacekeeper – the warrior peacekeeper, who employs conventional military skills, and the cosmopolitan peacekeeper that applies softer, less traditional skills to communicate and negotiate with warring factions and local people, implement community-engagement PoC tasks and support mission-specific peacebuilding efforts.[7] Yet, as Curran contends, improving peacekeeping training so that these divergent peacekeeper identities can be better reconciled to respond to complex demands in regions where 'war amongst the people' occurs remains a challenge.[8]

The UN defines peacekeeping training as 'any training activity which aims to enhance mandate implementation' by equipping uniformed

---

6   Sixty-five semi-structured interviews with RDF military personnel, trainers and external consultants were conducted during four field research trips conducted at Gako Military Training Academy in Musanze and the Ministry of Defence (MINEDEF) in Kigali, Rwanda between 2014 and 2015. Discourse analyses of depth-interviews and RDF training materials were triangulated with non-participatory observations of classroom-based training, a half-day observation of field exercises and desk-based research. All research participants consented to the interviews and are referred to by a number, rank and role to ensure anonymity. See Georgina Holmes, 'Situating agency, embodied practices and norm implementation in peacekeeping training', *International Peacekeeping* 76:4 (2018) for a detailed description of the methodology used in this study.

7   David Curran, *More than Fighting for Peace? Conflict, Resolution, UN Peacekeeping, and the Role of Training Military Personnel* (Cham: Springer, 2016), 119. For discussion on peacekeeper masculinities, see Paul Higate and Marsha Henry, *Insecure Spaces: Peacekeeping, Power and Performance in Haiti, Kosovo and Liberia* (London: Zed Books, 2009); Sandra Whitworth, *Men, Militarism and UN Peacekeeping: A Gendered Analysis* (London: Lynne Rienner, 2004); Claire Duncanson, 'Forces for Good? Narratives of Military Masculinity in Peacekeeping Operations', *International Feminist Journal of Politics* 11:1 (2009), 63–80; Marianne Bevan and Megan H. MacKenzie, '"Cowboy" Policing versus "the Softer Stuff"', *International Journal of Politics* 14:4 (2012), 508–38.

8   David Curran, *More than Fighting for Peace?*

personnel with the skills, capabilities and attitudes to 'meet the evolving challenges of peacekeeping operations, in accordance with DPKO/DFS principles, policies and guidelines, [and] lessons learnt from the field'.[9] General Assembly resolution A/RES/49/37 (1995) stipulates that TCCs are required to train uniformed personnel using CPTMs produced by UN DPKO/DFS and the Integrated Training Service (ITS). In this respect, the UN conceptualises pre-deployment training as a mechanism through which liberal peace norms, values, beliefs and technical skills are transferred to peacekeepers, rather than as a field of social practice wherein norms are translated and/or negotiated by actors engaged in the training process. In other words, an examination of how military leaders, military trainers, external consultants and individual trainees partake in the pre-deployment training programme is required to understand how norms are implemented in practice.

To date, few studies on peacekeeping training have accounted for the differences in experiences between men and women and the gender dynamics within the teaching environment[10] and the prevailing assumption in much of the existing research is that the soldier-trainee – or peacekeeper subject – is either gender-neutral or male.[11] Empirical studies exploring gender and peacekeeping training have tended to analyse gender training modules in isolation. These studies examine how male soldiers are trained and socialised as peacekeepers and fall into two categories: educating men about the purpose of gender mainstreaming and preventing Sexual Exploitation and Abuse (SEA) by enforcing norm compliance through behavioural change.[12] How female military peacekeepers learn soft skills and capabilities during the pre-deployment training programme, beyond combatant capabilities, and how they are trained to perform as

---

9 For details, see United Nations, *United Nations Peacekeeping Resource Hub* (Geneva: United Nations, 2016), available at http://research.un.org/en/peacekeeping-community/Training, accessed 28 October 2018.

10 See Sion, 'Peacekeeping and the Gender Regime'; Lindy Heinecken, 'Are Women 'Really' Making a Unique Contribution to Peacekeeping?', *Journal of International Peacekeeping* 19:3/4 (2015), 227–48.

11 See A.B. Featherston, 'UN Peacekeepers and Cultures of Violence', *Cultural Survival* 19:1 (1995); Nina Wilén and Lindy Heinecken, 'Regendering the South African Army: Inclusion, reversal and displacement', *Gender, Work and Organisation* 25:6 (2018), 1–17; Deborah Goodwin, *The Military and Negotiation: The Role of the Soldier-Diplomat* (London: Routledge, 2015); David Curran, 'Training for Peacekeeping: Towards Increased Understanding of Conflict Resolution', *International Peacekeeping* 20:1 (2013), 80–97.

12 See Paul Higate, *Military Masculinities: Identity and the State* (Westport: Praeger, 2003); Whitworth, *Men, Militarism and UN Peacekeeping*; Dean Laplonde, 'The Absence of Masculinity in Gender Training for UN Peacekeepers', *Journal of Social Justice* 27 (2015), 91–99; Lisa Carson, 'Pre-deployment "gender" training and the lack thereof for Australian peacekeepers', *Australian Journal of International Affairs* 70:3 (2016), 275–92.

peacekeepers has not been the subject of such studies. This is surprising, given that, increasingly, women are regarded as a 'gender resource', deployed to deliver specific tasks such as responding to rape survivors, as discussed below.

UNSCR 1325 and its related resolutions propose that gender mainstreaming should be systematically integrated into all systems and structures and across all security institutions operating in peacekeeping and peacebuilding contexts.[13] UN gender mainstreaming norms[14] purport that 'biological sex' and culturally constructed concepts of 'gender' should be separated to ensure that individuals, regardless of their sex, sexuality and gender, are empowered, treated fairly and provided equal opportunities within peacekeeping workforces.[15] Although the UN's gender mainstreaming norms derive from the UN Charter and are relatively stable, they are adaptive and adaptable and can be translated and operationalised in varied ways by different agents.[16] In this sense, the UN's gender mainstreaming norms can be broadened or narrowed once operationalised. When broadened, the norms are 'transformative', as they can set the policy agenda and transform existing organisational structures and ways of working.[17] When narrowed, gender mainstreaming norms become what is termed as 'integrationist', since they are 'submitted to goals' other 'than that of gender equality' and are made to 'fit into existing policy frames'.[18]

Conceptualising the Pre-Deployment Training Space

This part of the chapter draws on Bourdieu's theory of social practice to conceptualise the pre-deployment training space as a site of norm

---

13   United Nations, *United Nations landmark resolution on women, peace and security Resolution 1325* (New York: United Nations, 2000), available at http://www.un.org/womenwatch/osagi/wps/, accessed 28 October 2018.

14   Katzenstein defines norms as the 'collective expectations for the proper behaviour of actors with a given identity in global politics' that 'serve the purpose of guiding behaviour by providing motivations for action'. See Peter Katzenstein, *The Culture of National Security: Norms and Identity in World Politics* (New York: Columbia University Press, 1983), 5.

15   Marysia Zalewski, '"I Don't Even Know What Gender Is": A discussion of the connections between gender, gender mainstreaming and feminist theory', *Review of International Studies* 36:1 (2010), 12.

16   Jutta Joaquim and Andrea Schneiker, 'Changing discourses, changing practices? Gender mainstreaming and security', *Comparative European Politics* 10:5 (2012), 529; Mona L. Krook and Jackie True, 'Rethinking the life cycles of international norms: The United Nations and the global promotion of gender equality', *European Journal of International Relations* 18:1 (2010), 103–27.

17   Emilie Hafner-Burton and Mark A. Pollack, 'Mainstreaming Gender in Global Governance', *European Journal of International Relations* 10:3/4 (2002), 285–98.

18   Joaquim and Schneiker, 'Changing discourses, changing practices?', 530.

contestation.[19] All actors that engage in pre-deployment training are identified and located within a field and are both shaped by and shape the structures of the field. The interplay of socialised norms, practices and discourses that determine the behaviour and thinking of actors that operate within the field constitute the field's 'habitus'.[20] The pre-deployment training space is a hybrid space in which institutional actors from within the TCC military and external actors (such as consultants from training institutes, UN subsidiary groups, UN Women and DPKO/DFS) contribute to the operationalisation of the UN's gender mainstreaming norms. Since dominant control over the operationalisation of norms lies with the TCC military, actors – such as senior leaders or trainers – may seek to preserve their autonomy. For instance, TCC senior military leaders may reject outright UN gender mainstreaming norms and/or create informal (invisible) gender rules if they perceive that the military institution's established gender regime is threatened by the introduction of external gender norms. Thus, UN institutionalised norms are either accepted in their entirety, partially accepted or rejected outright by the TCC military's senior leaders and trainers, as well as external consultants, to make them 'workable and practical' in more than one local context – the TCC military and the peacekeeping operation.

The pre-deployment training space is a militarised field of power, but it is also a field of education wherein identities and roles (or subject positions) are constructed and negotiated during transfers of knowledge between educators and learners. Skeggs's political analysis of educational institutions helps to explain how the socialisation mechanisms used in pre-deployment training construct gendered peacekeeping subject positions through the training process. Informed by national and institutional discourses and practices, the learning curriculum helps to 'establish hierarchical relations between different forms of knowledge' and creates 'a network of subject positions in relation to these hierarchies'.[21] Subject positions are constructed according to institutionally and socially recognised (often dialectical) categorisations and 'cultural representations', such as 'man'/'woman', 'masculine'/'feminine', 'heterosexual'/'homosexual', 'black'/'white' and so on,[22] leading to the creation of 'scripted' peacekeeper subject positions. Scripted subject positions can be performed but may not

---

19 For a more detailed theorising of norm implementation in peacekeeping training, see Holmes, 'Situating agency, embodied practices and norm implementation in peacekeeping training'.

20 Pierre Bourdieu, *The Field of Cultural Production* (Cambridge: Polity Press, 1993), 30.

21 Beverley Skeggs, *Formations of Class and Gender: Becoming Respectable* (London: Sage Publications, 1997), 61.

22 Skeggs, *Formations of Class and Gender*, 12.

necessarily be internalised by individuals.[23] The socialisation processes embedded within training curricula and practices encourage men and women to accept their place within the sexual division of labour of the pre-deployment training space and consequently within the organisational structures of both the TCC military and the battalion deployed to the peacekeeping mission.[24]

Deploying Female Peacekeepers

The DPKO/DFS's first 'Gender Policy for UN Peacekeeping' (2010) and the DPKO/DFS 'Guidelines for Integrating a Gender Perspective into the work of the United Nations Military in Peacekeeping Operations' (2010) set out how the utility of female peacekeepers in cosmopolitan peacekeeping missions has become institutionalised in UN policy discourse in recent years. However, DPKO/DFS's implementation of the UN's gender mainstreaming norms is conflicting. First, DPKO/DFS adopt an integrationist (and instrumentalist) approach to gender mainstreaming by incorporating a gender perspective into established work streams, emphasising the requirement to increase female military personnel. This includes applying a gender perspective to PoC and law enforcement activities, disarmament, demobilisation and reintegration and security sector reform (DDR/SSR) programmes and Civil-Military (CIMIC) operations aligned to peacebuilding mission objectives. Female military peacekeepers are expected to extend the reach of peacebuilding mission objectives and engage with women and vulnerable groups when delivering capacity building projects, to empower local women and educate them about human rights and to act as liberated and empowered role models for local women, promoting democracy and freedom of speech.

Second, DPKO/DFS proposes that UN gender mainstreaming norms should be operationalised by transforming the structures of the mission itself, including the cultures and modes of operation within peacekeeping contingents, the military, police and civilian components and across multidimensional teams. Here, the DPKO/DFS interpret UN gender mainstreaming norms as transformative and serve the purpose of promoting equal opportunities for all members of the peacekeeping workforce. However, there exists an unwitting contradiction in how gender equality is expected to be mainstreamed. On one hand, women are expected to perform gender-specific roles as 'female peacekeepers'

---

23 Skeggs, *Formations of Class and Gender*, 12.
24 Skeggs, *Formations of Class and Gender*, 48.

to deliver mission objectives. On the other, DPKO/DFS promote the conscious separation of biological sex and gender to facilitate fair and equal treatment of all staff within security institutions operating within the peacekeeping system.

To support the implementation of gender mainstreaming norms, the UN gender directive advises that 'all modules for pre-deployment training for uniformed and civilian personnel [should] cover the role and rationale of work for gender equality and the empowerment of women in peacekeeping contexts'.[25] Although they are more progressive than the Standard Generalised Training Materials (SGTMs) that predated them, the CPTMs, published in 2009, promote an integrationist approach to gender mainstreaming by emphasising that a gender perspective should be incorporated into existing mission objectives to enhance operational effectiveness. Discussion on how gender mainstreaming norms can be operationalised to facilitate the transformation of ways of working within peacekeeping teams and contingents is absent. CPTM Unit 3 on implementing the Women, Peace and Security (WPS) agenda, as well as CPTM Unit 2, which concerns respecting cultural diversity, observe that 'national, institutional and professional differences' within and across military, police and civilian components exist. They do not teach military peacekeepers that gendered institutional norms and practices can prevent women from progressing to more senior decision-making positions within peacekeeping operations and potentially restrict their ability to work effectively to deliver mission mandates. The CPTMs were updated in 2017, with Module Two (WPS) of the 2017 CPTMs examining gender and power structures within the UN peacekeeping system in more detail. However, the question of how these training materials will be implemented by TCCs and external consultants is yet to be determined.

So far, the chapter has explained how the pre-deployment training space is a site of norm contestation as well as a military field of education and has examined how DPKO/DFS describe the utility of female peacekeepers in their gender policies. The following case study examines how the UN's gender mainstreaming norms are operationalised through the pre-deployment training programme for tactical-level Rwandan military peacekeepers deployed in a battalion and considers the implications this has for preparing Rwandan female military peacekeepers for their interactions with the local population in host communities experiencing protracted wars.

---

25 United Nations, *Core Pre-deployment Training Materials* (New York: United Nations, 2010), 16 available at http://repository.un.org/bitstream/handle/11176/89573/CPTM_All. pdf?sequence=1&isAllowed=y, accessed 28 October 2018.

Pre-Deployment Training in Rwanda

Pre-deployment training for Rwandan peacekeepers takes place at Gako Military Academy in Musanze, 25 km from the capital Kigali. Battalions tend to comprise 800 personnel, including one all-female unit of 10 to 20 personnel.[26] Peacekeepers spend two weeks of the three-month pre-deployment training programme in the classroom and six weeks engaged in field exercises. In the final month, they are deployed near their homes to allow them to make arrangements with their families and complete any outstanding administration prior to deployment to UN/African Union (AU) missions. Since RDF trainers deliver classroom-based modules, the UN's gender mainstreaming norms diffused through the DPKO's training materials were operationalised by internal norm brokers rather than external norm brokers (consultants). Although RDF Gender Desk trainers were tasked with teaching cadets basic training modules on gender-based violence and child protection, the CPTMs, including the gender module, were not taught by internal gender experts but by RDF officers assigned to train peacekeepers for deployment, since the TCC was responsible for delivering these courses. These men did not perceive the UN's more transformative gender mainstreaming norms to be relevant for preparing peacekeepers for deployment. The US-funded African Contingency Operations Training and Assistance (ACOTA) consultants offered technical advice on setting up practical exercises. However, they did not regard integrating gender into practical exercises as a priority and therefore did not offer external assistance to ensure that the UN's transformative gender mainstreaming norms were implemented across the pre-deployment training programme.[27]

Since the more complex understandings of gender mainstreaming that seek to transform peacekeeping workforces and promote equal opportunity peacekeeping were not considered to be of use in the theatre of operation, RDF trainers did not integrate gender issues across the pre-deployment training programme, as recommended by DPKO/DFS. Instead, they selected elements of the older SGTMs, which presented a narrower, integrationist reading of gender mainstreaming than the CPTMs. These older training materials were used in conjunction with modules developed by the AU and modules developed by the RDF. Taken together, this resulted in a hybridised pre-deployment training programme, where RDF ethics and values were taught to soldiers alongside UN/AU modes of operation, and RDF modes of operation were expected to be practised

---

26   Senior officer 1, interview with the author, 28 May 2014.
27   Brigadier General, interview with the author, 10 June 2014.

within Rwandan base camps within UNAMID. Peacekeepers received informational briefings on PoC, gender, human rights, child soldiers and humanitarian activities, as well as classroom-based trainings on effective crowd control, managing riots, demobilising armed groups and securing camps for Internally Displaced Persons (IDPs), amongst others. Indeed, the RDF only integrated gender into the modules on preventing SEA, peacekeeper discipline and performance, diversity awareness and PoC. As in most militaries, male and female personnel train together to ensure they deploy as a cohesive team. However, a Rwandan male peacekeeper subject was foregrounded in the training discourse and there was no conscious effort to consider the training needs of the female soldiers in the battalion. Instead, emphasis was placed on ensuring discipline among male troops.[28] Training objectives specified that the Rwandan male peacekeeper should understand the conceptual idea of gender as socially constructed, the different ways conflict affects men, women and children and the rationale for why male peacekeepers should not engage in SEA.[29] Yet, discussion of how gendered power relations in militaries and rules and regulations can facilitate gender inequalities in the workplace, however subtle, was not included.

Constructing the Scripted Rwandan Female Peacekeeper

RDF military personnel responses in interviews identified a scripted Rwandan female peacekeeper subject position. Women trainees regarded themselves as the same as their male colleagues when working in the Rwandan base camp, undertaking technical military tasks such as surveillance and camp security or, in three cases, repairing vehicles and communications equipment or working in administration. One major explained:

> For me, I go to the mission area as a peacekeeper, not a female. I do the task of S6, not S6 female. You know what? If I go there as a female peacekeeper, I go there like a female and I am not interested in my task. I do the tasks of a female. That's why I go there as a peacekeeper.[30]

Despite self-identifying as gender-neutral peacekeepers, two-thirds spoke of the need to perform a 'female peacekeeper' subject position to help deliver UNAMID mission objectives. One radio operator responded:

---

28 Female major, interview with the author, 4 June 2014.
29 Female peacekeeper 6, interview with author, 9 June 2015.
30 Female peacekeeper 35, interview with the author, 9 June 2014.

> Sometimes I act as a peacekeeper and other times as a female peacekeeper. For example, like those [issues] of gender. Sometimes we go and meet women and they ask us to give them things like bread, water, milk etc. If they see you are a woman, they come to you instead of men. So in that situation, I act as a female peacekeeper, although I am a peacekeeper in all other categories.[31]

The women interviewed explained that the Rwandan female peacekeeper was expected to engage with women from the local population within the mission and therefore required the skill set of a 'social worker' peacekeeper, which meant they should be caring, empathetic, able to listen to (predominantly) local women's concerns and be able to teach local women about their human rights. One private commented:

> The female peacekeeper must console women who have just been raped, and teach gender to the population. She must have empathy to console and to teach gender in Darfur. I will teach them about the gender factor and teach them [women] how to protect themselves. And I'll teach them to love the military.[32]

Many of the Rwandan female military peacekeepers interviewed did not regard the focus on discipline to prevent SEA in the modules on Code of Conduct, Rules of Engagement, Gender and RDF military ethics as being relevant to them. Rather, the women understood 'being disciplined' as performing as professional peacekeepers and ambassadors for post-genocide Rwanda. This meant ensuring that they upheld the core values of the RDF and performed as liberated, empowered, modern Rwandan women. Thus, the construction of the scripted Rwandan female peacekeeper was not just dependent on establishing a socially recognisable dialectic category of male/female peacekeeper, but on establishing a second dialectic category to distinguish between developed, empowered Rwandan women and underdeveloped, disempowered women-victims in the mission area. Here, the UN's emphasis on engaging tactical-level female military personnel in peacebuilding activities aligned well with national policy on the role of women as peacebuilders in post-conflict Rwanda.[33] The RDF localised the UN's integrationist gender mainstreaming norms so that the scripted Rwandan female peacekeeper performed as a Rwandan role model and educator (diffuser) of Rwandan norms and practices – notably

---

31 Female peacekeeper 8, interview with the author, 6 June 2015.
32 Female peacekeeper 27, interview with the author, 6 June 2014.
33 See Georgina Holmes, *Women and War in Rwanda: Gender, Media and the Representation of Genocide* (New York: I.B. Tauris, 2014); Holmes, 'Gendering the Rwanda Defence Force'; Holmes, 'Gender and the Military in post genocide Rwanda'.

related to conflict resolution and post-conflict development. According to one trainee, her 'participation as an African woman is to teach fellow African woman'.[34] Another remarked that she will take the women 'out of violence and teach them the benefits of unity and reconciliation, and try to teach them how to get out of ethnic conflict'.[35] In performing the scripted Rwandan female peacekeeper subject position, women were also required to perform as the bearer of Rwandan culture, for example, by wearing traditional dress during cultural exchanges.[36] In contrast, fewer men were expected to wear traditional Rwandan dress during cultural exchanges and only a small number joined dance troupes. Within UNAMID, tactical-level Rwandan female military personnel were expected to perform two subject positions: the gender-neutral peacekeeper engaged in tactical-level military tasks within the base camp and the disciplined Rwandan female peacekeeper, who cares for and educates local women-victims and embraces and promotes (RPF-sanctioned) traditional Rwandan culture and modern Rwandan values.[37]

Security Practices and Performances during Pre-Deployment Training

The rejection of the UN's transformative gender mainstreaming norms takes effect through norm subsidiarity and in the subtle operationalisation of the RDF's formal (codified and visible) and informal (invisible) gender rules that structure Rwandan base camps within UNAMID as well as the pre-deployment training space in Rwanda. For example, the RDF's gendered protection policy policed the sexual division of labour within the mission area and reproduced stereotypes that 'portrayed women as incapable of providing protection'.[38] In UNAMID, the majority of tactical-level female military peacekeepers worked solely within the Rwandan military base camp unless they were called upon to visit IDP camps. Women did not join male Rwandan military peacekeepers on short-duration day patrols unless it was known in advance that female military peacekeepers would be required[39] and they were not allowed to participate in night patrols or long-duration patrols. Only four out of the 37 women interviewed engaged in intelligence gathering.

---

34 Female peacekeeper 24, interview with the author, 5 June 2014.
35 Female peacekeeper 27, interview with the author, 6 June 2014.
36 Female peacekeeper 11, interview with the author, 10 June 2015.
37 Jennie E. Burnet, *Genocide Lives within the US: Women, Memory and Silence in Rwanda* (Wisconsin: University of Wisconsin Press, 2012); Holmes, *Women and War*.
38 Pruitt, *The Women in Blue Helmets*, 74.
39 Female peacekeeper 18, interview with the author, 13 June 2015; Female peacekeeper 16, interview with the author, 10 June 2010.

Task-orientated[40] practical exercises, designed to test peacekeepers on knowledge learnt in the classroom[41] addressed gender issues in narrow terms. During an observation of a training exercise, the battalion was presented with two scenarios. In the first, warring factions had set up roadblocks, were conducting ambushes and hindering peacekeepers from reaching IDP camps and isolated, dehydrated Sudanese women. In the second, the battalion had to undertake crowd dispersal following IDP demonstrations near the Rwandan base camp. During these scenarios, nine out of the ten women within the battalion undertook basic military tasks within the Rwanda base camp, which included providing security around the camp perimeter and at the gates, conducting body searches on women, manning the observation tower and assisting in communications operations. One woman – a captain – worked as an intelligence analyst. In the first scenario, female military personnel were expected to role-play as the scripted Rwandan female peacekeeper and tasks were assigned according to their gender. These process-orientated role-plays included teaching local women and girls about gender equality, searching the house of a poor woman, rebuilding houses and giving them water. Due to their low rank, the women were not expected to make decisions about managing convoys and ambushes. Neither male nor female peacekeepers were required to negotiate with female combatants during the patrol exercise because, according to senior RDF staff, based on their experience in UNAMID, it was considered 'rare to find women in ambushes in Darfur'.[42]

The acceptance of the integrationist gender mainstreaming norms, whereby women were valued for their ability to undertake specific gendered tasks, such as engaging with women from the local population, appeared to more closely align to the existing RDF (informal) gender norms and thus could be more easily translated into practice in the two local contexts within which tactical-level military personnel are required to operate: UNAMID, the mission where the peacekeepers were deployed, and the TCC's military institution.

Assessment of Pre-Deployment Training by Female Peacekeepers

Overall, female military trainees regarded the pre-deployment training they received more positively than women who had returned from the mission. However, most of the women interviewed spoke of the training related to

---

40 RDF trainer 2, interview with the author, 5 June 2014.
41 RDF trainer 1, interview with the author, 11 December 2014.
42 Senior RDF staff 8, interview with author, 7 June 2014.

military tasks, rather than the soft skills required for PoC activities. When asked, many felt confident they would be able to engage effectively with traumatised people in complex situations because, as Rwandan women who had grown up in a post-conflict state, they would be able to empathise with them, regardless of the training provided. A 2nd lieutenant explained, 'we have experience. For example, the genocide in our country – we saw many things. This made us able to help other countries overcome their war problems'.[43] There was an assumption that, because of their shared gender, local women would easily open up and communicate with them, regardless of their age, sexuality, dis/ability and personal circumstances:

> *Private*: For example, she has been raped. She can't talk to a man, but she feels free to talk to me.
>
> *Interviewer*: Have you been specifically trained on how to talk to a woman that has experienced conflict-related sexual violence?
>
> *Private*: I have not received that training, but as a person I can imagine what it must be like to speak to her … What is happening in Darfur is almost the same as what happened in Rwanda because they kill each other and [the women and children] are left as widows and orphans. So I really pity them because I know what they are going through and I feel I should counsel them.[44]

Several degree-educated women awaiting deployment were more critical of the pre-deployment training programme. They felt the training on gender issues was inadequate because it was too short, did not go into depth about how conflict affected men and women differently[45] and focused on norm compliance and behavioural change, rather than on how to respond to gender issues as tactical-level military peacekeepers. One trainee remarked, 'I don't call it training, it's more like an information briefing because they teach us how to behave in the mission and to perform like men'.[46]

Of those that had returned from missions, women who operated only in their professional military trades within the base camp or mission headquarters and were not required to engage directly with the local population in challenging contexts considered their pre-deployment training to be adequate. In contrast, women who were engaged in complex protection duties or were required to work in isolated teams outside of the Rwandan base camp wanted to enhance their communication and

43 Female peacekeeper 41, interview with the author, 18 November 2015.
44 Female peacekeeper 29 interview with the author, 6 June 2014.
45 Female peacekeeper 29 interview with the author, 6 June 2014.
46 Female peacekeeper 25, interview with the author, 5 June 2014.

negotiation skills so that they could better handle difficult situations. Dealing with local people who were traumatised was the main concern discussed. A major who had deployed to UNAMID as a liaison officer felt that training peacekeepers to follow a short protocol for assisting rape survivors did not prepare them for prolonged engagements with survivors over several hours. Referring to the pre-deployment training schedule, she remarked:

> All these things were helpful. But when I encountered gender-based violence [in the mission] it was not easy ... when I was on the ground and reached the [IDP] camp, you find this person who was raped for two or three hours and the people [around her] don't want to communicate, they don't care about what happened. So you take care of her. You don't go alone, you go with military observers and there must be a female there, the police – they must be female. So you get this person, you put her in touch with the NGOs, you take her to hospital, but you need to spend three or four hours with her.[47]

Male peacekeepers also lacked the training to handle traumatised local people including men, since training referring to gender issues focused on 'the conduct of Rwandan soldiers' and preventing SEA.[48] However, returned female peacekeepers rejected the idea that soft skills, such as communication, negotiation and empathy, were feminine traits. One private requested techniques on how to have 'conversations with local people in order to find out where the problem is and how they really live'.[49] The Liaison Officer wanted more information on the 'psychological impact' conflict-related sexual violence has on survivors.[50] In parallel with the need for deeper context on gender issues and soft skills, several women requested more mission-specific scenarios and interactive role-plays for PoC activities. A female captain requested training on how to 'better implement a gender perspective into the conduct of operations' in intelligence analysis and was keen to learn how 'gender could be integrated into decision-making'. She believed more complicated scenarios taught in smaller classes would allow her to learn how peacekeepers 'handle problems and execute decisions well' in more complex situations such as in 'wars amongst the people'.[51]

These examples demonstrate how operationalising the UN's gender mainstreaming norms in narrow terms limits the ways in which female military peacekeepers are trained to handle complex situations that they

---

47   Female peacekeeper 2, interview with the author, 3 June 2014.
48   Male trainer 6, interview with the author, 9 June 2014.
49   Female peacekeeper 11, interview with the author, 10 June 2015.
50   Female peacekeeper 2, interview with the author, 3 June 2014.
51   Female peacekeeper 23, interview with the author, 5 June 2014.

experience when providing security in 'wars amongst the people'. While the socialisation of Rwandan female military peacekeepers ensures they understand their basic roles in the mission, the formula becomes problematic for those women who are required to perform 'off-script', as they so often do in real life scenarios.

Conclusion

The UN considers pre-deployment training to be an important mechanism by which to transfer the security institutions' gender mainstreaming norms to male and female peacekeepers. It has been argued that the pre-deployment training space is a distinct site of norm contestation and negotiation. The analysis presented here shows that the TCC military's expectation and allocation of (gendered) resources in the mission determines how tactical-level female military peacekeepers are trained and socialised. In the Rwanda case study, the RDF partially conformed to UN gender mainstreaming norms because the military supported an integrationist approach, rather than a combined integrationist/transformative approach to gender mainstreaming. The latter approach would require that peacekeepers were taught that gender mainstreaming has the potential to change the structures, policies and social practices (or modes of operation) within the mission area and the TCC military.

This analysis does not suggest that, during missions, women did not act dynamically, flexibly or creatively. Rather, it demonstrates how the RDF constructs a scripted Rwandan female peacekeeper subject position based on how senior officers perceive women's value within the mission. The structural constraints of the pre-deployment training programme reinforce normative gender stereotypes that position female peacekeepers as naturally possessing the soft skills required to negotiate, communicate and empathise with local populations in the host country. The findings of the research reaffirm the requirement to facilitate active learning through increasingly thematic-based scenarios, role-plays and simulations, allocating more time for peacekeepers to learn the soft skills required for community-engagement/PoC tasks. The UN is increasingly providing funding and support to develop training programmes for female officers prior to their deployment to UN peacekeeping operations. Yet, for TCCs, such as Rwanda, who are required to introduce many topics in a short space of time during the classroom-based trainings, there is a requirement to think creatively about how training can be transformative and how best to utilise active learning methods to allow peacekeepers trained in mixed-gender battalions to engage with some of the more challenging issues the UN's gender mainstreaming norms seek to address. Integrating a deeper study of gender issues into the existing pre-deployment training programme will not solve what is ostensibly a political and structural problem. While UN gender advisors could support the design and implementation of TCC-led pre-deployment training programmes, gender mainstreaming needs to be addressed at the strategic level within TCC militaries and by national governments to ensure effective gendered security sector reform within national security institutions.

References

Beardsley, K. and Karim, S., 'Female peacekeepers and gender balancing: Token gestures or informed policymaking?', *International Interactions* 39:4 (2013).

Bevan, M. and MacKenzie, M., '"Cowboy" Policing versus "the Softer Stuff"', *International Journal of Politics* 14:4 (2012).

Björkdahl, A. and Selimovic, J., 'Translating UNSCR 1325 from the global to the local: protection, representation and participation in the National Action Plans of Bosnia-Herzegovina and Rwanda', *Conflict, Security and Development* 14:4 (2015).

Bourdieu, P., *The Field of Cultural Production* (Cambridge: Polity Press, 1993).

Bridges, D. and Horsfall, D., 'Increasing Operational Effectiveness in UN Peacekeeping: Toward a Gender-Balanced Force', *Armed Forces and Society* 36:1 (2009).

Burnet, J., *Genocide Lives within the US: Women, Memory and Silence in Rwanda* (Wisconsin: University of Wisconsin, 2012).

Carson, L., 'Pre-deployment 'gender' training and the lack thereof for Australian peacekeepers', *Australian Journal of International Affairs* 70:3 (2016).

Curran, D., 'Training for Peacekeeping: Towards Increased Understanding of Conflict Resolution', *International Peacekeeping* 20:1 (2013).

Curran, D., *More than Fighting for Peace? Conflict, Resolution, UN Peacekeeping, and the Role of Training Military Personnel* (Cham: Springer, 2016).

Davies, S. and True, J. (eds), *The Oxford Handbook of Women, Peace and Security* (Oxford: Oxford University Press, 2019).

De Groot, G.J., 'A few good women: Gender stereotypes, the military and peacekeeping', *International Peacekeeping* 8:2 (2001).

Duncanson, C., 'Forces for Good? Narratives of Military Masculinity in Peacekeeping Operations', *International Feminist Journal of Politics* 11:1 (2009).

Elliot, L. and Cheeseman, G. (eds), *Forces for Good: Cosmopolitan Militaries in the Twenty-first Century* (Manchester: Manchester University Press, 2004).

Featherston, A.B., 'UN Peacekeepers and Cultures of Violence', *Cultural Survival* 19:1 (1995).

Goodwin, D., *The Military and Negotiation: The Role of the Soldier-Diplomat* (London: Routledge, 2015).

Hafner-Burton, E. and Pollack, M., 'Mainstreaming Gender in Global Governance', *European Journal of International Relations* 10:3/4 (2002).

Heathcote, G. and Otto, D. (eds), *Rethinking Peacekeeping, Gender Equality and Collective Security* (London: Palgrave Macmillan, 2014).

Heinecken, L., 'Are Women 'Really' Making a Unique Contribution to Peacekeeping?', *Journal of International Peacekeeping* 19:3/4 (2015).

Higate, P., *Military Masculinities: Identity and the State* (Westport: Praeger, 2003).

Higate, P. and Henry, M., *Insecure Spaces: Peacekeeping, Power and Performance in Haiti, Kosovo and Liberia* (London: Zed Books, 2009).

Holmes, G., *Women and War in Rwanda: Gender, Media and the Representation of Genocide* (New York: I.B. Tauris, 2014).

Holmes, G., 'Situating agency, embodied practices and norm implementation in peacekeeping training', *International Peacekeeping* 76:4 (2018).

Holmes, G., 'Gender and the Military in Post-Genocide Rwanda' in Elissa Bemporad and Joyce W. Warren (eds), *Women and Genocide: Survivors, Victims, Perpetrators* (Bloomington: Indiana University Press, 2018).

Joaquim, J. and Schneiker, A., 'Changing discourses, changing practices? Gender mainstreaming and security', *Comparative European Politics* 10:5 (2012).

Karim, S. and Beardsley, K., *Equal Opportunity Peacekeeping* (Oxford: Oxford University Press, 2017).

Katzenstein, P., *The Culture of National Security: Norms and Identity in World Politics* (New York: Columbia University Press, 1983).

Krook, M. and True, J., 'Rethinking the life cycles of international norms: The United Nations and the global promotion of gender equality', *European Journal of International Relations* 18:1 (2010).

Laplonde, D., 'The Absence of Masculinity in Gender Training for UN Peacekeepers', *Journal of Social Justice* 27 (2015).

Olsson, L. and Tryggestad, T., *Women and International Peacekeeping* (London/New York: Routledge, 2001).

Pruitt, L., *The Women in Blue Helmets: Gender, Policing and the UN's First All-Female Peacekeeping Unit* (California: University of California Press, 2016).

Rupesinghe, N., Karlsrud, J. and Stamnes, E., 'Women, Peace and Security and Female Peacekeeping Personnel' in S. Davies and J. True (eds), *The Oxford Handbook of Women, Peace and Security* (Oxford: Oxford University Press, 2019).

Simic, O., 'Increasing Women's Presence in Peacekeeping Operations: The Rationales and Realities of 'Gender Balance' in G. Heathcote and D. Otto (eds), *Rethinking Peacekeeping, Gender Equality and Collective Security* (London: Palgrave Macmillan, 2014).

Sion, L., 'Peacekeeping and the Gender Regime: Dutch female peacekeepers in Bosnia and Kosovo', *Journal of Contemporary Ethnography* 37:5 (2008).

Skeggs, B., *Formations of Class and Gender: Becoming Respectable* (London: Sage Publications, 1997).

United Nations, *Core Pre-deployment Training Materials* (New York: United Nations, 2010) 16, available at http://repository.un.org/bitstream/handle/

11176/89573/CPTM_All.pdf?sequence=1&isAllowed=y, accessed 28 October 2018.

United Nations, *United Nations landmark resolution on women, peace and security Resolution 1325* (New York: United Nations, 2000), available at http://www.un.org/womenwatch/osagi/wps/, accessed 28 October 2018.

United Nations, *United Nations Peacekeeping Resource Hub* (Geneva: United Nations, 2016), available at http://research.un.org/en/peacekeeping-community/Training, accessed 28 October 2018.

Whitworth, S., *Men, Militarism and UN Peacekeeping: A Gendered Analysis* (London: Lynne Rienner, 2004).

Wilén, N. and Heinecken, L., 'Regendering the South African Army: Inclusion, reversal and displacement', *Gender, Work and Organisation* 25:6 (2018).

Zalewski, M., '"I Don't Even Know What Gender Is": A discussion of the connections between gender, gender mainstreaming and feminist theory', *Review of International Studies* 36:1 (2010).

PART THREE
LEGAL DEBATES

7

REDEFINING THE 'COMBATANT'

Just War Theory and 'Direct Participation in Hostilities'

Andree-Anne Melancon

Introduction

The Law of Armed Conflict (LOAC) is generally presented as having four basic underlying principles: humanity, military necessity, proportionality and distinction.[1] This last requirement is especially important when discussing 'war amongst the people' as it refers to the distinction between civilians and combatants. This is of such importance that the International Committee of the Red Cross (ICRC) recognises it is the first rule of Customary International Humanitarian Law (CIHL): 'Rule 1. The parties to the conflict must at all times distinguish between civilians and combatants. Attacks may only be directed against combatants. Attacks must not be directed against civilians'.[2] In this case, 'combatants' is understood to refer to members of the regular armed forces. As the ICRC states 'Rule 3. All members of the armed forces of a party to the conflict are combatants, except medical and religious personnel'.[3] Further, the principle of distinction was codified in the 1977 *Additional Protocol I*:

---

1   See, for example, LOAC Blog, '4 Basic Principles', *LOAC Blog* October 2015, available at https://loacblog.com/loac-basics/4-basic-principles/, accessed 28 October 2018.

2   International Committee of the Red Cross, 'Rule 1. The Principle of Distinction between Civilians and Combatants', *IHL Database*, available at https://ihl-databases.icrc.org/customary-ihl/eng/docs/v1_rul_rule1, accessed 28 October 2018.

3   International Committee of the Red Cross, 'Rule 3. Definition of Combatants', *IHL Database*, available at https://ihl-databases.icrc.org/customary-ihl/eng/docs/v1_rul_rule3, accessed 29 October 2018.

Article 48 – Basic rule:
In order to ensure respect for and protection of the civilian population and civilian objects, the Parties to the conflict shall at all times distinguish between the civilian population and combatants and between civilian objects and military objectives and accordingly shall direct their operations only against military objectives.[4]

Although the application of this distinction requirement can appear straightforward, this is far from the case, especially in contemporary conflict. Situations of asymmetrical warfare, counterinsurgency and counterterrorism blur the line between civilians and combatants, with the rising importance, for example, of 'part-time combatants' or 'farmer-fighters' (described as farmers by day and fighters by night). As the ICRC states:

One of the key elements of international humanitarian law is the clear distinction between members of the armed forces and civilians. In contemporary armed conflicts, however, the proximity of civilians to military operations and their increased assumption of traditionally military functions lead to confusion as to the implementation of the principle of distinction.[5]

It is here that the notion of 'direct participation in hostilities' (DPH) takes its importance. The concept of 'participation in hostilities' has evolved from Common Article 3 of the 1949 Geneva Conventions calling for the humane treatment of 'Persons taking no active part in the hostilities'.[6] This led to a more precise definition in the 1977 Additional Protocols. Paragraph 3 of Article 51 of the first Protocol states that 'Civilians shall enjoy the protection afforded by this Section, unless and for such time as they take a direct part in hostilities'.[7] The same exception is found in Article 13 of the second Protocol, which notes that 'Civilians shall enjoy the protection afforded by this Part, unless and for such time as they take a direct part in hostilities'.[8] Thus, the concept of DPH is an exception or a limit to civilian

---

4   International Committee of the Red Cross, *Protocol Additional to the Geneva Conventions of 12 August 1949 and Relating to the Protection of Victims of International Armed Conflicts (Protocol I)* (Geneva: International Committee of the Red Cross, 1977), 3.

5   International Committee of the Red Cross, 'Civilian 'Direct Participation in Hostilities': Overview', *ICRC* October 2010, available at https://www.icrc.org/eng/war-and-law/contemporary-challenges-for-ihl/participation-hostilities/overview-direct-participation.htm, accessed 29 October 2018.

6   The full text of all four Geneva Conventions can be accessed online at https://ihl-databases.icrc.org/applic/ihl/ihl.nsf/vwTreaties1949.xsp accessed 29 October 2018.

7   International Committee of the Red Cross, *Protocol Additional to the Geneva Conventions of 12 August 1949 and Relating to the Protection of Victims of International Armed Conflicts.*

8   International Committee of the Red Cross, *Protocol Additional to the Geneva Conventions of 12 August 1949 and Relating to the Protection of Victims of Non-International Armed Conflicts (Protocol II)* (Geneva: International Committee of the Red Cross, June 1977), 609.

immunity in conflict. Yet, even considering the importance of DPH, the concept is not defined. Indeed, the two Additional Protocols do not expand on the definition or meaning of 'direct participation' beyond the statements presented earlier. Thus, as the ICRC has observed, 'The contemporary challenge, therefore, is to provide clear criteria for the distinction not just between civilians and the armed forces, but also between peaceful civilians and civilians who directly participate in hostilities'.[9]

In the wider context of exploring 'war amongst the people', this chapter will focus on the interpretation of DPH using the just war tradition. It begins with an overview of distinction or discrimination in the just war literature. In light of this, a novel approach to the principle of discrimination will be presented. The chapter will then argue that this new approach can be used to clarify what is understood by DPH, making distinction easier to apply in the context of 'war amongst the people'.

The Just War Tradition

There is no single universal approach to just war theory. As Ian Clark has pointed out:

> It is not possible to speak of a single doctrine of just war, nor can we point to the lineal development of a single idea; nor can we talk of the doctrine having a continuous history. At best, just war doctrine is a set of recurrent issues and themes in the discussion of warfare and reflects a general philosophical orientation towards the subject.[10]

In its review of the morality of warfare, the just war tradition generally focuses on two aspects of the use of force: *jus ad bellum* (how force is *justified*) and *jus in bello* (how force is *used*).[11] In addition to this, a number of just war theorists, championed by Brian Orend, add *jus post bellum* (justice after the conflict) in their interpretation of just war. This refers to the just conclusion of the war, including the peace process and reparations if necessary.[12] Discussions on 'war amongst the people' are set in the *jus in bello* framework

---

9 International Committee of the Red Cross, "Civilian 'Direct Participation in Hostilities'": Overview'.

10 Ian Clark, *Waging War: A Philosophical Introduction* (Oxford: Oxford University Press, 1988), 31.

11 Colm McKeogh, *Innocent Civilians: The Morality of Killing in War* (New York: Palgrave Macmillan, 2002), 4.

12 Brian Orend, 'War' in Edward N. Zalta (ed.), *The Stanford Encyclopedia of Philosophy Archives* (Stanford: Stanford University Press, 2005), available at https://plato.stanford.edu/archives/spr2016/entries/war/, accessed 28 October 2018.

and refer to issues related to the principle of discrimination (or distinction) that calls for the use of force to be aimed at legitimate targets only.

Before the discussion can turn to just war theory and the contribution it presents when exploring 'war amongst the people', it is important to present a brief overview of the debate surrounding the link between *jus in bello* and *jus ad bellum*. This is important because the way the link is understood has an impact on the interpretation of discrimination or distinction. If *jus in bello* and *jus ad bellum* are independent one from the other, then unjust wars (those that do not fulfil the *jus ad bellum* requirements) can still be fought justly (following *jus in bello* rules), meaning that all combatants are equal, regardless of what side they are on. This is known as the moral equality of combatants. However, some authors argue that unjust wars can never be fought justly because combatants on the unjust side lack just cause and thus the harm they cause cannot be justified. Broadly, there are two 'families' in the just war tradition: the orthodox (or traditional) and the revisionist. Traditional / orthodox just war thinking is rooted in collectivism or the 'collective nature of warfare',[13] while revisionist just war thinkers reject collectivism in favour of an individualist approach, based on the view of war as being constituted of a multitude of individual acts.[14]

A key figure in orthodox or traditional just war thinking is Michael Walzer, author of the influential text *Just and Unjust Wars*. Walzer's argument on war is rooted in the independence of *jus ad bellum* and *jus in bello*: 'In our judgments of the fighting, we abstract from all consideration of the justice of the cause. We do this because the moral status of individual soldiers on both sides is very much the same: they are led to fight by their loyalty to their own states and by their lawful obedience'.[15] Building on this, Walzer's approach to the discrimination criterion is based on collectivism and focuses on interstate conflicts. He states that 'the first principle of the war convention is that, once war has begun, soldiers are subject to attack at any time (unless they are wounded or captured)'.[16] His approach to *jus in bello* discrimination is grounded in the moral equality of combatants, which he calls 'battlefield equality'.[17] For Walzer, all members of the soldier 'class' (group) are legitimate targets because 'simply by fighting, whatever their

---

13 Seth Lazar, 'Just War Theory: Revisionists vs Traditionalists', 3. Available at https:// static1.squarespace.com/static/56301fcfe4b0b0dc3b73f211/t/5771d7419de4bbb95cf214 8a/1467078466322/ARPS+War+v1-4.pdf accessed 28 October 2018.

14 This is an over-simplification of the debates between orthodox and revisionist just war thinkers, but is sufficient to provide context for the discussion of this section exploring the just war tradition and 'war amongst the people'.

15 Michael Walzer, *Just and Unjust Wars: A Moral Argument with Historical Illustrations* (London: Basic Books, 2006), 127.

16 Walzer, *Just and Unjust Wars*, 138.

17 Walzer, *Just and Unjust Wars*, 137.

private hopes and intentions, they have lost their title to life and liberty'.[18] Moreover, civilians that are not part of the soldier class are illegitimate targets, even if they are citizens of a state engaged in an unjust war: 'The war convention rests first on a certain view of combatants, which stipulates their battlefield equality'.[19] Essentially, Walzer's account of discrimination/ distinction in traditional just war theory echoes the approach to distinction found in LOAC; both rely on group membership to identify legitimate targets. This collective approach to the interpretation of discrimination implies that even non-threatening combatants can be considered legitimate targets regardless of their broad involvement in the fighting (because all members of the soldier class are equal). To exemplify this, Walzer uses the case of the 'naked soldier', arguing that a combatant taking a bath, while being non-threatening, remains a legitimate target because he remains engaged in the collective activity of war. Walzer argues that 'it is not against the rules of war as we currently understand them to kill soldiers who look funny, who are taking a bath, holding up their pants, revelling in the sun, smoking a cigarette'.[20] In addition to all soldiers being legitimate targets, Walzer's second war convention argues that 'noncombatants cannot be attacked at any time. They can never be the objects or targets of military activity'.[21] This is how non-combatant immunity is commonly understood. However, this immunity is not absolute: for instance, munitions factory workers can be attacked in war because of their contribution to the war effort.[22] They are assimilated to the soldier class whilst they are in their workplace and return to the civilian class when they leave the factory. Munitions factory workers are also differentiated from individuals working in a factory providing food or medicine for the army:

> The relevant distinction is not between who works for the war effort and those who do not, but between those who make what soldiers need to fight and those who make what they need to live, like the rest of us. When it is militarily necessary, workers in a tank factory can be attacked and killed, but not workers in a food processing plant.[23]

In short, Walzer's discrimination criterion is membership-based and differentiates between combatants (soldiers) and non-combatants (civilians). When exploring 'war amongst the people' and the notion of

---

18 Walzer, *Just and Unjust Wars*, 136.
19 Walzer, *Just and Unjust Wars*, 137.
20 Walzer, *Just and Unjust Wars*, 142.
21 Walzer, *Just and Unjust Wars*, 151.
22 Walzer, *Just and Unjust Wars*, 145.
23 Walzer, *Just and Unjust Wars*, 146.

DPH, this approach is pertinent because it recognises that certain civilians can also be legitimate targets in conflict, having been assimilated into the soldier class because their actions are considered participation in the hostilities.

This traditional/orthodox approach to the *jus in bello* principle of discrimination bears many similarities with the one set out in international law. It is founded on the belief in the collective nature of warfare bringing a membership-based approach to the identification of legitimate targets in warfare, where soldiers are legitimate targets and civilians are illegitimate targets unless they contribute to the war effort. This presents interesting insight when it comes to interpreting DPH, as some acts can be considered as direct participation if they contribute to the war effort.

Just as it is impossible to discuss a single unitary just war theory, there is no single account of revisionist just war thinking. For the sake of this chapter however, Jeff McMahan's account is useful as it formed the foundation of much of the revisionist 'movement'. McMahan's conception of the morality of war is in opposition to the one presented by Walzer and other traditional/orthodox just war theorists. Fundamentally, McMahan rejects the moral equality of combatant and the independence of *jus ad bellum* and *jus in bello*. This means that combatants in an unjust war according to *jus ad bellum* are not morally justified in fighting: 'it is morally wrong to fight in a war that is unjust because it lacks just cause'.[24] McMahan argues that 'if just combatants were always to fight according to the moral constraints that govern their conduct in war, they would never be liable to attack and unjust combatants would never have legitimate targets at all and thus would never satisfy the requirement of discrimination'.[25] Moreover, he rejects the moral equality of combatants and collectivism, viewing it as being irreconcilable with reality outside the context of war or conflict.[26] In that sense, in times of peace, the use of force is asymmetrical: the victim of an attack can use self-defence against the aggressor, but the aggressor is not entitled to self-defence against the victim. Yet, belief in battlefield equality implies that both the attacker and the victims are equally entitled to the use of force. More broadly, McMahan argues that, if a different set of *moral* criteria apply in war as opposed to everyday life, then some clear criteria are necessary in order to differentiate between times of war and times of peace and between competing sets of moral rules.[27] In the absence of this clear distinction, the rules of war ought to also apply to everyday life and vice versa.

---

24 Jeff McMahan, *Killing in War* (Oxford: Oxford University Press, 2009), 6.
25 McMahan, *Killing in War*, 6–7.
26 McMahan, *Killing in War*, 35.
27 McMahan, *Killing in War*, 35–36.

With regard to the discrimination between legitimate and illegitimate targets, McMahan draws attention to the existence of two different notions of 'combatants'.[28] Firstly, there is the legal conception of combatant, as codified in international law and presented earlier in this chapter. Secondly, there is a moral conception of combatant, which relies on moral responsibility: 'a person may be morally liable to having such force used against him simply by virtue of being morally responsible for the existence of an unjust threat [lacking just cause], even if he does not himself pose the threat'.[29] McMahan's discrimination criterion departs from membership-based liability as suggested by Walzer.[30] This is also a rejection of the collective conceptions of warfare, with the focus resting on individuals and their actions. The implication of this is that civilians can be moral combatants: 'civilians can abet the prosecution of an unjust war. And they can be complicit in instigating it' either by supporting the conflict or failing to oppose it.[31] He also highlights the importance of the *jus in bello* proportionality of means requirements and states:

> Most civilians have, on their own, no capacity at all to affect the actions of their government. [Thus] military attack exceeds what a person may ordinarily be liable to on the basis of these comparatively trivial sources of responsibility. This is to say, in effect, that most international attacks on civilians violate the narrow proportionality requirement that is internal to the notion of liability.[32]

Briefly, McMahan's liability and discrimination criterion rely on individual moral responsibility for unjust threats. Responsibility is interesting in the context of interpreting the notion of DPH since it moves beyond the membership-based approach found in traditional just war thinking and international law. When exploring 'war amongst the people', revisionist approaches to discrimination suggest an individual-based approach, which is also favoured when interpreting DPH.

ICRC Guidance

Through the twentieth century, war moved away from conventional battlefields and was increasingly fought 'amongst the people'. The intervention in Afghanistan consolidated this trend. In 2003, the ICRC set

---

28   McMahan, *Killing in War*, 11–13.
29   McMahan, *Killing in War*, 10.
30   Jeff McMahan, 'Liability and Collective Identity: A Response to Walzer', *Philosophia* 34 (2006), 15.
31   McMahan, *Killing in War*, 214.
32   McMahan, *Killing in War*, 225.

up an 'Expert Process' to clarify the concept of DPH and explore three key questions:

- Who is considered a civilian for the purpose of conducting hostilities and, therefore, must be protected against direct attack 'unless and for such time as they directly participate in hostilities'?
- What conduct amounts to direct participation in hostilities and, therefore, suspends a civilian's protection against direct attack?
- What modalities govern the loss of civilian protection against direct attack (including its duration; the precautions and presumptions to be observed in situations of doubt; the restraints imposed on the use of force against lawful targets and the consequences of restoring civilian protection)?[33]

Between 2003 and 2008, five meetings were organised, gathering 40–50 experts from military, academic, governmental and non-governmental backgrounds, to explore the ICRC's questions and to clarify the concept of DPH. However, the fifth and final Expert Meeting held in February 2008 ended without unanimous consent.[34] This lack of consensus even amongst experts illustrates the challenges associated with defining and refining DPH. Nevertheless, in 2009, the ICRC published its *Interpretive Guidance on the Notion of Direct Participation in Hostilities under International Humanitarian Law*. The *Guidance* is based on the discussions and debates that occurred during these meetings, as well as further research conducted by the ICRC.

The starting point of the ICRC's approach to DPH is to establish it as a specific act: 'The notion of direct participation in hostilities refers to specific acts carried out by individuals as part of the conduct of hostilities between parties to an armed conflict'.[35] The ICRC identifies three essential constitutive elements of DPH: a threshold of harm, a causal link and a belligerent nexus. The first constitutive element of the notion of DPH can be summarised as follows: 'In order to reach the required threshold of harm, a specific act must be likely to adversely affect the military operations or military capacity of a party to an armed conflict, or alternatively, to inflict death, injury or destruction on persons or objects protected against direct attack'.[36] The key element of this definition is 'military harm' and can

---

33 Nils Melzer, *Interpretative Guidance on the Notion of Direct Participation in Hostilities under International Humanitarian Law* (Geneva: International Committee of the Red Cross, 2009), 13.
34 This is why no list of participants was published. See Melzer, *Interpretative Guidance on the Notion of Direct Participation in Hostilities*, 4.
35 Melzer, *Interpretative Guidance on the Notion of Direct Participation in Hostilities*, 45.
36 Melzer, *Interpretative Guidance on the Notion of Direct Participation in Hostilities*, 47.

include electronic interference or denying the enemy the use of certain objects or equipment.[37] Such acts are considered as DPH because they are the equivalent to methods used in warfare. Acts that do not cause military harm (not military in nature), such as building a fence or interrupting food supplies, do not meet the harm threshold for DPH.[38] In addition to this, the infliction of death or injury to protected persons are considered 'adverse effect', thereby meeting the harm threshold. Therefore, an example of an act constituting DPH would be a direct attack against civilians or civilian objects – a common occurrence in 'war amongst the people'.

The second constitutive element of DPH is direct causation. Essentially, 'there must be a direct causal link between a specific act and the harm likely to result either from that act or from coordinated military operation of which that act constitutes an integral part'.[39] The guidance draws the line between direct and indirect causation as being 'one causal step'.[40] Indirect actions (involving more than one causal step) include recruiting and scientific research. These acts are considered as being too removed from the actual harm to constitute DPH. Interestingly, the ICRC's document indicates a lack of consensus on the causality of certain actions, such as voluntary human shields.[41]

The third constitutive element of DPH is the belligerent nexus. This means that 'an act must be specifically designed to directly cause the required threshold of harm in support of a party to the conflict and to the detriment of another'.[42] This element refers to the 'participation' aspect of DPH and is not to be conflated with 'hostile intent', which only 'relates to the state of mind of the person concerned, whereas belligerent nexus relates to the objective purpose of the act. The purpose is expressed in the design of the act or operation and does not depend on the mindset of every participating individual'.[43] Essentially, this constitutive element distinguishes between activities that are considered DPH and activities that might still bring some harm to protected persons but are not related to the hostilities that are occurring within the wider context of the conflict, such as civil unrest.[44]

Finally, with regard to the temporal aspect of DPH, the *Interpretative Guidance* states that 'civilians lose protection against direct attack for

---

37 Melzer, *Interpretative Guidance on the Notion of Direct Participation in Hostilities*, 48.
38 Melzer, *Interpretative Guidance on the Notion of Direct Participation in Hostilities*, 50.
39 Melzer, *Interpretative Guidance on the Notion of Direct Participation in Hostilities*, 51.
40 Melzer, *Interpretative Guidance on the Notion of Direct Participation in Hostilities*, 52–53.
41 Melzer, *Interpretative Guidance on the Notion of Direct Participation in Hostilities*, 57.
42 Melzer, *Interpretative Guidance on the Notion of Direct Participation in Hostilities*, 58.
43 Melzer, *Interpretative Guidance on the Notion of Direct Participation in Hostilities*, 59.
44 Melzer, *Interpretative Guidance on the Notion of Direct Participation in Hostilities*, 64.

the duration of each specific act amounting to direct participation in hostilities'.[45] This 'revolving door' approach does not apply to members of the regular forces (combatants) who can be targeted as long as they remain in their continuous combat function and are not *hors de combat*.

Unfortunately, the ICRC's guidance is not without its shortcomings. Schmitt is correct in his analysis that, although the three constitutive elements identified in the *Interpretive Guidance* are important, they are 'under-inclusive'.[46] They fail to acknowledge that DPH can occur beyond the 'one step' level of causation. Indeed, one side's military capacity can be considerably enhanced by actions that fall outside this limit, such as recruitment activities. The key implication of the *Guidance*'s limit is that it can lead states ultimately to reject or ignore the ICRC's interpretation of DPH. This can be seen through the practice of targeting individuals when they are not directly participating in hostilities, as was the case with the drone strike that killed Anwar al-Awaki in Yemen.[47] The case of improvised explosive device (IED) makers illustrates the disagreement between state practice and the ICRC's *Guidance*. On the topic of causation, the ICRC states: 'the assembly and storing of an improvised explosive device in a workshop, or the purchase or smuggling of its components, may be connected with the resulting harm through an uninterrupted causal chain of events, but, unlike the planting and detonation of that device, do not cause harm directly'.[48] Thus, the ICRC distinguishes between IED makers and those who plant the device. This bears many similarities with the case of munitions factory workers who are not considered to be taking a direct part in the hostilities: 'such work, occurring far from the battlefield and with no knowledge of how, where and when the ammunition will be used, is not integral to any particular operations or group of operations'.[49] However, this interpretation is not universally accepted. The causality of IED fabrication was debated during the 2006 Fourth Expert Meeting[50] and few would consider IED assembly as being too far removed to constitute DPH since it could fall into the 'hostile intent' category often used as part of rules of engagement (RoE). As Schmitt observed, experts were divided on this issue: 'Nearly all those with military experience or who serve

---

45 Melzer, *Interpretative Guidance on the Notion of Direct Participation in Hostilities*, 70.

46 Michael N. Schmitt, 'Deconstructing Direct Participation in Hostilities: The Constitutive Elements', *International Law and Politics* 42 (2010), 739.

47 For a more detailed review of the acceptance of the ICRC's guidance, see Ka Lok Yip, 'The ICRC's interpretive guidance on the notion of direct participation in hostilities: sociological and democratic legitimacy in domestic legal orders', *Transnational Legal Theory* 8:2 (2017), 224–46.

48 Melzer, *Interpretative Guidance on the Notion of Direct Participation in Hostilities*, 54.

49 Schmitt, 'Deconstructing Direct Participation in Hostilities', 731.

50 Melzer, *Interpretative Guidance on the Notion of Direct Participation in Hostilities*, 50–51.

governments involved in combat supported the characterization of IED assembly as direct participation'.[51] Another example of a contentious case relating to DPH has to do with the transport of weapons. The *Guidance* states:

> The delivery by a civilian truck driver of ammunition to an active firing position at the front line would almost certainly have to be regarded as an integral part of ongoing combat operations and, therefore, as direct participations in hostilities. Transporting ammunition from a factory to a port for further shipping to a storehouse in a conflict zone, on the other hand, is too remote from the use of that ammunition in specific military operations to cause the ensuing harm directly.[52]

Although these two examples might not appear problematic, the ICRC does not explore the more troublesome and problematic situations between driving to the front line and driving towards the fight. Following the 'one causal step' approach, only driving to the front line would constitute DPH. However, in the 2008 *Strugar* appeals case, the International Tribunal for the Prosecution of Persons Responsible for Serious Violations of International Humanitarian Law Committed in the Territory of the Former Yugoslavia since 1991 (ICTY) took a broader approach to this issue and included 'transporting weapons in proximity to combat operations'[53] in its list of examples of DPH. This type of transport could easily fall outside of the 'one causal step' limit set by the ICRC. Thus, there remains a level of practical ambiguity as to the meaning of 'direct' with regard to causation in DPH.

In sum, even with the ICRC's efforts to provide *Interpretative Guidance*, the notion of DPH remains open to interpretation. Importantly, the causal and temporal limitation of the ICRC's approach means that states can and do establish their own practice and interpret DPH in a way that is favourable to their operations, in part because the ICRC is seen as too restrictive. This permits a relaxation of the rules, stretching what is permissible according to International Humanitarian Law (IHL) and, as a consequence, an increasingly large number of actions are seen as being indicative of combatant status, as is the case with the use of US 'signature strikes' in drone warfare.

---

51 See note 98 in Schmitt, 'Deconstructing Direct Participation in Hostilities', 731.

52 Melzer, *Interpretative Guidance on the Notion of Direct Participation in Hostilities*, 56.

53 United Nations International Tribunal for the Prosecution of Persons Responsible for Serious Violations of International Humanitarian Law Committed in the Territory of the Former Yugoslavia since 1991, *Prosecutor v Srugar Case No. IT-01-42-A, Appeals Judgement* (Geneva: United Nations, July 2008), para 177, available at http://www.icty.org/x/cases/strugar/acjug/en/080717.pdf, accessed 28 October 2018.

Applying Discrimination

Considering the interpretations of discrimination and DPH presented above, this section will focus on developing an approach to discrimination that can be applied in contemporary conflict where war occurs 'amongst the people'.

In just war tradition, the tension between revisionist and traditional/ orthodox also concerns issues of applicability. In that sense, revisionist thinking is often described as being moral philosophy and can be criticised as being inapplicable in the realities of conflict. It is not hard to see how evaluating individual degrees of moral responsibility and support for the war effort for the entire civilian population in the midst of battle can be difficult, if not impossible. This is similar to certain issues that arise with the notion of DPH, as there is an assumption that combatants can easily identify the types of acts that constitute 'direct participation', thereby fulfilling the ICRC's requirements. In an exchange on the moral equality of combatants with Walzer, McMahan agreed that his interpretation of the discrimination criterion is difficult to apply. Furthermore, McMahan's concern lies in the 'deep morality of war'.[54] On the topic of McMahan's interpretation of discrimination, Walzer states: 'What he actually provides, I think, is a careful and precise account of what individual responsibility in war would be like if war were a peacetime activity'.[55] Here, Walzer's criticism bears many similarities with the real-world objection as formulated by Strawser:

> I may grant that the revisionist objections to the traditional just war theory's tenets (such as the rejection of the M[oral] E[quality of] C[ombatants]) are technically correct, and I grant that the revisionist account for liability to be killed in war is morally superior. But be that as it may, there's just no way to implement the revisionist account in the real world, so it must be rejected.[56]

Applicability and concerns such as the real-world objection are of great importance if a workable understanding of civilian participation in hostilities is to be reached. The rejection of revisionist (and individualist) approaches to discrimination raises another issue, highlighted by Lazar: 'By arguing that liability is grounded in responsibility for unjustified threats,

---

54   Jeff McMahan, 'The Ethics of Killing in War', *Philosophia* 34 (2006), 23–41.

55   Michael Walzer, 'Response to McMahan's Paper', *Philosophia* 34 (2006), 43.

56   Bradley J. Strawser, 'Revisionist Just War Theory and the Real World: A Cautiously Optimistic Proposal' in Fritz Allhoff, Nicholas G. Evans and Adam Henschke (eds), *Routledge Handbook of Ethics and War* (Basingstoke: Routledge, 2013), 78. A full version of the chapter is available at https://www.academia.edu/10149233/Revisionist_Just_War_Theory_and_the_Real_World_A_Cautiously_Optimistic_Proposal, accessed 28 October 2018.

not the fact that one poses the threat, McMahan opens the floodgates to total war'.[57] This then leads to a rather difficult dilemma:

> If we raise the liability threshold too high, then many who we think ought to be legitimate targets in war will be excluded. On the other hand, if we lower the liability threshold in response to this problem, then it invites its own problem: far too many people will fall into the category of being legitimate targets than is plausible for a just war.[58]

The danger of total war in this sense is a reality when 'war amongst the people' occurs. It bears many similarities with contemporary counterterrorism and asymmetric warfare, where an increasingly large proportion of populations can be considered legitimate targets, even with the considerable technological and technical advances now available.

It is important to reiterate that policy and morality are different and must not be conflated. For example, 'non-combatant immunity' is often understood as being synonymous with 'discrimination between legitimate and illegitimate targets' but they are not interchangeable. Non-combatant immunity is the regulatory principle whereas the discrimination between legitimate and illegitimate targets is the moral principle. The key implication of this is the acknowledgement that morality and applied ethics can be multi-levelled. Indeed, on the metaphorical morality ladder, policy principles are at the bottom and moral principles are located above the policy, at different levels depending on the approach. If the requirement (liability threshold) is too high, then the requirement is too difficult to apply, as articulated in the real-world objection. The danger here is that, because of this lack of real-life applicability, the moral principle is ignored in practice. This then brings another danger where the liability threshold is set too low and too many individuals become considered legitimate targets, moving towards a total war. To avoid both of these situations, a secondary theory is necessary. A *juste milieu* between ideal revisionist just war theory and the current status quo is required. To do this, discrimination will be interpreted as being two-fold. Firstly, discrimination is understood in a way that sets the guiding moral principle for targeting in conflict. However, this is insufficient to make war more just and more discriminate. In addition to this, a secondary interpretation of discrimination is necessary to make the criterion more applicable in everyday conflict. This is essential to make contemporary asymmetrical warfare more just and for the discrimination criterion to be implemented and upheld in conflict.

---

57 Seth Lazar, 'The Responsibility Dilemma for *Killing in War*: A Review Essay', *Philosophy and Public Affairs* 38:2 (2010), 187–88.

58 Strawser, 'Revisionist Just War Theory and the Real World', 78.

Individual Discrimination

Firstly, the *jus in bello* discrimination criterion is understood as being similar to the high morality of war. The foundation of this interpretation is the revisionist account of the just war tradition and the importance of individual rights. The battlefield equality is rejected, so too is the idea that there can be a clear line between legitimate and illegitimate targets. Instead, liability is based on individual moral responsibility for threats in war, as suggested by revisionist just war thinking. This is *individual discrimination*. At this level, three different types of individuals are legitimate targets. Firstly, being the commander or the leader of a group of fighters brings individual moral responsibility for threats in the conflict. Without the power and decision-making of commanders and leaders, the fighting would be considerably diminished. Additionally, these individuals are generally involved in the organisation of the group and its decision-making. Secondly, known and identified recruiters can be considered legitimate targets because they are an essential factor in the creation of the threat in the conflict. Thirdly, key facilitators can become legitimate targets if they provide what is essential to the fighting, which makes them individually morally responsible for threats. For example, a bomb maker can be considered morally responsible for threats in the conflict and thus a legitimate target but the wife of a commander who provides food and shelter for her husband cannot and thus is not a legitimate target. Of course, even considering the moral validity of this approach, the applicability problem remains. There will be cases where the information necessary to identify commanders, leaders, recruiters or key facilitators will not be available, explaining the need for a secondary approach to discrimination. Nevertheless, individual discrimination is not impossible although it is difficult, as it sets a high liability threshold. There are cases where individual moral responsibility has been used to identify legitimate targets. The most recent example of this would be the use of 'kill lists', such as the American 'disposition matrix', or 'most wanted lists' in counterterrorism and asymmetrical warfare. There will be cases when individual moral responsibility for threats is impossible to establish, yet the individuals *should* be considered legitimate targets because they are posing an imminent threat as a result of their actions. This is where a second interpretation of discrimination becomes necessary.

Situational Discrimination

In addition to individual discrimination, a more concessive interpretation is needed. This acts as a supplement to the morality of war, in order to make it more applicable in conflict and to avoid the real-world objection.

Here, discrimination is dependent on the situations in which individuals find themselves. Legitimate targets are the ones who are engaged in harm or who pose an immediate threat. As Orend argues: 'A legitimate target in wartime is anyone or anything engaged in harming. All non-harming persons, or institutions, are thus ethically and legally immune from direct and intentional attack by soldiers and their weapon systems'.[59] This approach allows for some flexibility in the identification of targets as it simplifies the decision-making process, as information on responsibility is not necessary. Further, situational discrimination offers greater applicability than individual discrimination, in addition to being temporally flexible. Indeed, an important underlying implication of situational discrimination is that, since some individuals may be legitimate targets because of the situations they are in, they are not *always* legitimate targets. This also accounts for the fact that individuals are not consistently involved in the conflict, with some contributing sporadically or only once. For example, individuals whose identities are unknown (meaning that their individual moral responsibility for threats following individual discrimination cannot be established) can become a legitimate target because of the situations in which they are engaged. If they are fabricating explosives, they pose an immediate threat and lethal force can be used to neutralise the threat if there are no other available options, such as sending ground troops to capture the bomb makers.

It is important to remember that discrimination is not the only *jus in bello* requirement; proportionality also plays a key role. Indeed, assigning liability and identifying legitimate targets are not sufficient to make the use of force just. Indeed, force also ought to be proportional to the military objective sought. Thus, an individual might bear a low degree of liability that would make a target legitimate, but only to some harm, not to be killed. An example of this might be an individual who frequently expresses support for the conflict and contributes financially to the 'war effort' without taking part in the fighting or being a key facilitator. This liability threshold is especially important in contemporary conflict, with technological advances such as drones where these weapons are limited to the use of lethal force and capture is impossible. The implication of this is that, in applying the principle of discrimination, there ought to be a greater awareness of the weapons that will be used. Considering the central role of airpower in contemporary military deployment and the difficulty of assigning varying degrees of liability, focusing on differentiating between targets liable to be killed and those who cannot be legitimately killed, makes the application of

---

59   Brian Orend, *The Morality of War* (Peterborough: Broadview Press, 2006), 107.

discrimination more realistically possible. The use of lethal force still needs to be proportional to the military objective sought.

Briefly, the *jus in bello* principle of discrimination can and should be seen as being multi-layered. Firstly, legitimate targets are identified as such if they are individually morally responsible for threats following individual discrimination. In addition to this, situational discrimination adds greater applicability to the criterion, since it accounts for the targeting of targets of opportunity whose responsibility is unknown. In situational discrimination, legitimate targets are identified as such because they pose an immediate threat. These two approaches to discrimination can now be used to revisit the challenge of understanding DPH in the context of 'war amongst the people'.

Revising DPH

As noted earlier, even with the ICRC's *Interpretative Guidance,* debates on the scope of DPH remain. Nevertheless, it is possible to overcome some of these issues by revisiting the identification of legitimate targets in the just war literature. The framework combining individual and situational discrimination presented earlier is of great utility when it comes to the notion of DPH. Firstly, individual discrimination identifies legitimate targets for their responsible contributions to threats in the conflict. This is a move away from membership-based approaches to combatant status, as presented in IHL. It is important to eschew a focus on group membership in the identification of legitimate targets because this is ill-adapted to the realities of contemporary conflict, especially in cases of 'war amongst the people'. Civilians who bear some moral responsibility for threats, by being a key recruiter for example, are legitimate targets and are directly participating in hostilities even if they are not taking part in the fighting directly. Secondly, situational discrimination identifies legitimate targets as individuals who pose an immediate threat. Any civilian falling into this category directly participates in hostilities. Combined, individual and situational discrimination cover the different types of 'combatants' and individuals that can be encountered in conflict and offer some clear advantages.

Firstly, this new interpretation of discrimination/distinction allows for more rapid applicability and ease of understanding, in large part due to the similarities between self-defence and situational discrimination. This understanding of discrimination is less restrictive than the ICRC's interpretation in terms of causality. Returning to the issue of IED assembly, the *Guidance* argues that such action is causally too far removed to constitute DPH. However, this should not be the case; the 'one causal step' is too restrictive and leads each state to establish their own interpretation of DPH in their RoE. Instead of limiting 'direct participation' to one causal step away from the harm itself, DPH should include any actions judged essential to the fighting such as, *inter alia*, the fabrication of explosives, the funnelling of funds on the black market, transporting weapons, recruitment and propaganda. This is because the fighting would not happen without such actions and thus, they should be sufficient to make one a legitimate target in conflict. Finally, another advantage of the interpretation of discrimination presented earlier is that it clarifies the temporal scope of DPH. If targets are identified as being legitimate following individual discrimination (because their actions bring an individual responsibility for the threats in the conflict), they can be targeted at all times, unless there is clear evidence to suggest that they are no longer involved in the conflict by expressing the intent to surrender for example. This means that a civilian IED maker can be targeted even when not actively engaged in the assembly of explosive devices. In addition to this, situational discrimination offers a more limited temporal scope of application since it is more context-dependent. Individuals will only be legitimate targets for the time they pose an immediate threat when their level of responsibility is unknown (and thus individual discrimination cannot be applied). An example of this would be the transporting of explosives in the back of a truck near conflict, but not to the front lines. The ICRC's *Guidance* considers this too causally removed to be DPH but the ICTY disagreed. It can be difficult to determine the level of responsibility of the driver because he may not know that some of the cargo he is carrying includes weapons destined to be used in the conflict. In this case, the driver is an innocent threat but a threat nonetheless, which can be identified as a legitimate target with situational discrimination. Here, the temporal scope of DPH is limited to the driving.

In short, using individual and situational discrimination allows for a more flexible approach to the identification of legitimate targets, which then makes the interpretation of DPH less problematic. This is because legitimate targets are identified as such because of their individual responsibility or because of threats they pose, instead of relying on status.

Conclusion

A key issue that arises when 'war amongst the people' occurs has to do with the identification of legitimate targets. The requirement of distinction or discrimination is a fundamental rule of *jus in bello*. It is also a rather complicated issue, especially in contemporary warfare where the line between civilians and combatants is blurred to an even greater extent. The notion of DPH exists to account for civilian involvement in conflict but remains a challenge to define. This is why a move beyond combatant status and group membership would facilitate the identification of legitimate targets. Using the two-fold approach to discrimination presented here would overcome the issues related to distinction and more specifically to the definition of DPH. Indeed, using individual discrimination to identify legitimate targets on the grounds of an individual moral responsibility for threats allows for the targeting of civilians who directly participate in hostilities whilst also overcoming the issues linked to the ICRC's 'one causal step' interpretation. In addition to this, individuals posing an immediate threat or currently engaged in harm are directly participating in hostilities and are legitimate targets, regardless of status. Ultimately, this could make the use of force 'amongst the people' more discriminate.

References

Clark, I., *Waging War: A Philosophical Introduction* (Oxford: Oxford University Press, 1988).

Geneva Conventions, online at https://ihl-databases.icrc.org/applic/ihl/ihl.nsf/vwTreaties1949.xsp accessed 29 October 2018.

International Committee of the Red Cross, 'Rule 1. The Principle of Distinction between Civilians and Combatants', *IHL Database*, available at https://ihl-databases.icrc.org/customary-ihl/eng/docs/v1_rul_rule1, accessed 28 October 2018.

International Committee of the Red Cross, 'Rule 3. Definition of Combatants', *IHL Database*, available at https://ihl-databases.icrc.org/customary-ihl/eng/docs/v1_rul_rule3, accessed 29 October 2018.

International Committee of the Red Cross, *Protocol Additional to the Geneva Conventions of 12 August 1949 and Relating to the Protection of Victims of International Armed Conflicts (Protocol I)* (Geneva: International Committee of the Red Cross, June 1977).

International Committee of the Red Cross, *Protocol Additional to the Geneva Conventions of 12 August 1949 and Relating to the Protection of Victims of Non-International Armed Conflicts (Protocol II)* (Geneva: International Committee of the Red Cross, 1977).

International Committee of the Red Cross, 'Civilian 'Direct Participation in Hostilities': Overview', *ICRC* 29 October 2010, available at https://www.icrc.org/eng/war-and-law/contemporary-challenges-for-ihl/participation-hostilities/overview-direct-participation.htm, accessed 29 October 2018.

Lazar, S., 'Just War Theory: Revisionists vs Traditionalists', available at https://static1.squarespace.com/static/56301fcfe4b0b0dc3b73f211/t/5771d7419de4bb95cf2148a/1467078466322/ARPS+War+v1-4.pdf, accessed 28 October 2018.

Lazar, S., 'The Responsibility Dilemma for *Killing in War*: A Review Essay', *Philosophy and Public Affairs* 38:2 (2010).

McKeogh, C., *Innocent Civilians: The Morality of Killing in War* (New York: Palgrave Macmillan, 2002).

McMahan, J., 'Liability and Collective Identity: A Response to Walzer', *Philosophia* 34 (2006).

McMahan, J., 'The Ethics of Killing in War', *Philosophia* 34 (2006).

McMahan, J., *Killing in War* (Oxford: Oxford University Press, 2009).

Melzer, N., *Interpretative Guidance on the Notion of Direct Participation in Hostilities under International Humanitarian Law* (Geneva: International Committee of the Red Cross, 2009).

Orend, B., 'War' in E.N. Zalta (ed.), *The Stanford Encyclopedia of Philosophy Archives* (Stanford: Stanford University Press, 2005), available at https://plato.stanford.edu/archives/spr2016/entries/war/, accessed 28 October 2018.

Orend, B., *The Morality of War* (Peterborough: Broadview Press, 2006).

Schmitt, M.N., 'Deconstructing Direct Participation in Hostilities: The Constitutive Elements', *International Law and Politics* 42 (2010).

Strawser, B.J., 'Revisionist Just War Theory and the Real World: A Cautiously Optimistic Proposal' in F. Allhoff, N.G. Evans and A. Henschke (eds), *Routledge Handbook of Ethics and War* (Basingstoke: Routledge, 2013). A full version of the chapter is available at https://www.academia.edu/10149233/Revisionist_Just_War_Theory_and_the_Real_World_A_Cautiously_Optimistic_Proposal, accessed 28 October 2018.

United Nations International Tribunal for the Prosecution of Persons Responsible for Serious Violations of International Humanitarian Law Committed in the Territory of the Former Yugoslavia since 1991, *Prosecutor v Srugar Case No. IT-01-42-A, Appeals Judgement* (Geneva: United Nations, July 2008), Para 177, available at http://www.icty.org/x/cases/strugar/acjug/en/080717.pdf, accessed 28 October 2018.

Walzer, M., *Just and Unjust Wars: A Moral Argument with Historical Illustrations* (London: Basic Books, 2006).

Walzer, M., 'Response to McMahan's Paper', *Philosophia* 34 (2006).

Yip, K.L., 'The ICRC's interpretive guidance on the notion of direct participation in hostilities: sociological and democratic legitimacy in domestic legal orders', *Transnational Legal Theory* 8:2 (2017).

8

REVOLVING DOORS AND HUMAN RIGHTS

Detention in Internationalised Non-International Armed Conflict

Grant Davies

Introduction

A fragile state (the host state) invites the armed forces from another state (the assisting state) to help stabilise the nation in its struggle against an insurgency. High intensity conflict is prevalent, with loss of life on both sides. An insurgent is captured by the assisting state, but the commander is aware that the insurgent may be questioned only briefly before he must be transferred to the host state for prosecution. Due to the nature of combat operations, very little evidence has been collected that would assist any prosecution.

Such was the common experience of many commanders in recent conflicts, notably in Afghanistan and Iraq, both of which could be described as 'wars amongst the people'. In such situations, the fragile state's legal system will likely lack literate policemen, trained judges and prosecutors; judges and police are habitually intimidated, poorly paid and corruption rife; the political leadership may refuse to use administrative internment within its borders, as this would alienate it politically. The judiciary, police and prison services are usually grossly underdeveloped, and detainee abuse common. In summary, the chances of insurgents being convicted are minimal and the likelihood of them re-joining the fight worryingly high. The outcome, in short, is a revolving door for insurgents. Moreover, recent conflicts have shown that, at home, litigation will rage against the assisting state regarding detention, with severe financial and reputational cost.[1]

How then can the assisting state effectively detain in a Non-International Armed Conflict (NIAC) without incurring such risks? To answer this, several questions must be addressed. What rules govern detention in which type(s) of conflict? What is the legal basis for detention by the assisting state and why can the assisting state not simply retain detainees itself? What prevents the assisting state from simply transferring them to the assisted state? If a detainee must be transferred at all, where are the rules governing this? Why, during such conflicts, does international human rights law (IHRL) apply and not the traditional laws of war (International Humanitarian Law[2] (IHL)?[3] Should both regimes apply, and if so, which legal framework takes precedence?

This chapter adopts the perspective of an assisting state within a fragile host state.[4] It is presumed that no domestic law within the fragile state exists that would allow foreign states to detain, as is most often the case. The legality of such intervention (the *jus ad bellum*) will not be discussed, as the applicable law within the conflict (the *jus in bello*) is unaffected by it. Three critical issues form the basis of the discussion: the grounds for/basis of detention, detainee status and the modalities of detention in a NIAC. Relatively uncontroversial, the physical treatment of detainees will not be discussed.[5]

The term 'detainee' is not defined in any IHL treaty, nor is it applied consistently in practice.[6] This chapter will consider detention in two forms:

---

1 In *Al-Jedda*, the applicant was awarded damages and costs of €65,000. This does not include costs borne by the state in its own defence. For details, see European Court of Human Rights, *Al-Jedda v The United Kingdom App. No. 27021/08* (Strasbourg: ECtHR, 7 July 2011), available at http://www.bailii.org/eu/cases/ECHR/2011/1092.html, accessed 29 October 2018.

2 This is also known as the Laws of War or the Law of Armed Conflict (LOAC). This chapter will use the term IHL to encompass all.

3 A thorough synopsis of detention issues can be found in John B. Bellinger III and Vijay M. Padmanabhan, 'Detention Operations in Contemporary Conflicts: Four Challenges for the Geneva Conventions and Other Existing Law', *American Journal of International Law* 105:2 (2011), 201.

4 For studies of non-state actors see Sandesh Sivakumaran, *The Law of Non-International Armed Conflict* (Oxford: Oxford University Press, 2012), 236–41; Andrew Clapham, 'Human Rights Obligations of Non-State Actors', *International Review of the Red Cross* 88:86 (2006), 491–522; Sandesh Sivakumaran, 'Binding Armed Opposition Groups', *International and Comparative Law Quarterly* 55:2 (2006), 369–94.

5 For an extensive study into this, see Nigel Rodley and Matt Pollard, *The Treatment of Prisoners under International Law* (Oxford: Oxford University Press, 2009).

6 Gary Solis, *The Law of Armed Conflict* (Cambridge: Cambridge University Press, 2010), 244–45; Danish Ministry of Defence, *The Copenhagen Process on the Handling of Detainees in Military Operations* (Copenhagen: Danish Ministry of Defence, 2012), 1, paras i–vi.

'security internees' (internment without trial for an indefinite period) and 'criminal detainees' (detention as part of a criminal investigation). Only detention for reasons connected to the armed conflict is considered here. Legal regimes such as IHRL, IHL, refugee law and domestic law (of both the assisting state and the fragile state) are inextricably linked in NIAC and it is impossible to discuss them without referring to each other.[7]

Detention in 'Wars Amongst the People'

Public International Law recognises two forms of armed conflict: NIAC: conflict between a state and non-state actors and International Armed Conflict (IAC): armed conflict between the armed forces of two or more states. The former has, by far, been the most prevalent form of armed conflict since 1949.[8] Commonly referred to as 'insurgencies', such conflicts take place in the villages and towns of places such as Lashkar Gar and Basra, not between massed armies in the open fields of Kursk or the plains of El Alamein. These wars are truly 'wars amongst the people'. Accordingly, examination of the rules within them has never been more important.

Another recent trend is that third party states have also increasingly become involved in NIACs on behalf of the fragile state, thus 'internationalising' the 'wars amongst the people' in Mali, Nigeria, Afghanistan, Iraq and Syria and elsewhere. Unhelpfully, the amount of substantive law within IHL in NIAC is minimal, nowhere more so than in the law regarding detention. Compounding the problem, IHL risks being superseded by IHRL in NIAC to fill perceived legal gaps where clear but exhaustive rules can be found regarding detention, for instance concerning the Right to Liberty.

The chapter begins by exploring why detention is critical for assisting states and the option of simply 'not doing detention' is militarily sub-

---

7 For example, on 13 December 2006, the Public Committee against Torture in Israel v the Government of Israel (the 'Targeted Killings' case) drew upon Israeli domestic law, IHL and IHRL regarding the issue of 'kill or capture' in armed conflict. See Louise Doswald-Beck and Jean-Marie Henckaerts (eds), *International Humanitarian Law Volume I: Rules* (Cambridge: Cambridge University Press, 2005), xxvii; See also the International Committee of the Red Cross, *Expert meeting on procedural safeguards for security detention in non-international armed conflict* (London: ICRC, 2009), available at https://www.icrc.org/eng/assets/files/other/irrc-876-expert-meeting.pdf, accessed 29 October 2018.

8 United Nations, *The Secretary-General's High-level Panel Report on Threats, Challenges and Change: A more secure world, our shared responsibility* (New York: United Nations, 2004), available at https://www.un.org/ruleoflaw/files/gaA.59.565_En.pdf, accessed 29 October 2018.

optimal. Focus will then turn to the legal basis for detention, the status of the captured person and the modalities of detention. A final section outlines and considers future challenges and possible ways ahead.

Why Detain At All?

States face numerous legal and reputational challenges regarding detention operations,[9] as part of what some commentators have described as 'lawfare'.[10] Given such risks, the simple option would be to decide not to detain at all. In 2013, this rationale informed France's actions in Mali, where it had joined a NIAC in support of the government there. However, for a number of reasons, including lack of host nation capacity, logistical capability and 'reach', this decision was reversed. Still cognisant of and quickly confronted by the risks, the French forces in Mali found they had no option but to begin to detain insurgents.[11]

From one perspective, there are significant benefits to detention. In IAC, the taking of Prisoners of War (POW) 'yields many benefits … it reduces the enemy's numerical strength … their fighting capacity, and morale'.[12] Such benefits also occur in NIAC, in support of the stabilisation effort and to enhance force protection. Detention also symbolically supports the legitimacy of the fragile state being assisted – the total control of a person is a potent symbol of sovereignty, statehood and control.

One of the most compelling reasons to detain is that detainees are a quintessential source of human intelligence – an essential commodity in counterinsurgency operations. In recent campaigns, such intelligence

---

9   Although not an exhaustive list, it is worth considering the following cases: *Amnesty International v Canada Case* 2008 FC 336 (12 March 2008); *Hamdan v Rumsfeld*, 548 US 557 126 S.Ct 2749 (29 June 2006) (*Hamdan*); *Al-Skeini and Others v The United Kingdom*, App. No. 55721/07 (ECtHR 7 July 2011); *Ghousouallah Tarin v Ministry of Defence* (Denmark), Case no. B1-627-07.

10 For details on the term 'Lawfare', see Robert Chesney, Jack Goldsmith and Benjamin Wittes, 'About lawfare: A brief history of the term and the site', *Lawfare Blog*, September 2010, available at https://www.lawfareblog.com/about-lawfare-brief-history-term-and-site, accessed 29 October 2018. This describes 'Lawfare' as 'the strategy of using – or misusing – law as a substitute for traditional military means to achieve an operational objective'. Further explanation can also be found at https://thelawfareproject.org, accessed 29 October 2018.

11 This was a point made by Capt. Joris Cuzin, a French Legal officer with the French UN forces in Mali during a presentation to the UK Army Legal Services Annual Conference on 20 June 2013.

12 Development, Concepts and Doctrine Centre, *UK Joint Defence Publication 1-10 on Captured Personnel* (London: Ministry of Defence, 2015), available at https://assets.publishing.service.gov.uk/government/uploads/system/uploads/attachment_data/file/455589/20150820-JDP_1_10_Ed_3_Ch_1_Secured.pdf, accessed 29 October 2018.

was felt to have been the driver for the vast majority of operations. Furthermore, some detainees can be, and ultimately are, released and continue to provide intelligence – an advantage denied if insurgents are killed in the conflict. In short, while detention cannot by itself counter an insurgency, without it such a campaign can be lost. However, although detention in NIAC entails significant legal and political risk, it is far too valuable a tool to surrender and, in many cases, simply cannot be avoided without significant loss of military capability.

The Classification of Conflicts

As previously discussed, international law recognises two types of conflict – IAC and NIAC – under Common Articles 2 and 3 respectively (CA2 and CA3) of the Geneva Conventions (GC).[13] Rightly, much attention has been paid to the categorisation of conflicts[14] and, although the homogenisation of many of the rules has occurred, some vital differences remain, particularly regarding detention.[15] CA3 defines NIAC as those conflicts '*not* of an international character occurring in the territory of one of the High Contracting Parties' (emphasis added). NIAC was therefore envisioned as occurring *within one state*, between the state authorities and non-state entities; in effect, a 'classic' civil war.[16] However, even when an assisting state intervenes on the part of the fragile state, as outlined above, the conflict would still generally be recognised as a NIAC.[17] Furthermore, and critically, unless the threshold of 'armed conflict' has been met, IHL will not apply at all and the state's domestic law (including applicable IHRL) will apply exclusively. Unfortunately, no objective authoritative body exists to adjudicate on the classification and states often wish to preserve sovereignty by utilising their own laws and emergency powers,

---

13 These are called 'Common Articles' as these articles are common to all four of the Geneva Conventions.

14 An extensive study into classification of conflicts can be found in Dapo Akande, 'Classification of Armed Conflicts: Relevant Legal concepts' in Elizabeth Wilmshurst (ed.), *International Law and the Classification of Conflicts* (Oxford: Oxford University Press, 2012).

15 Although based on Customary Law Study No 6, not all states agree with the methodology or conclusions. See Sivakumaran, *The Law of Non-International Armed Conflict*, 61.

16 For a synopsis of the older regime of 'belligerency' see Anthony Cullen, 'Key Developments affecting the Scope of Internal Armed Conflict in International Humanitarian Law', *Military Law Review* 183 (2005), 65.

17 See *The Prosecutor v Jean-Pierre Bemba*, ICC-01/05-01/08-424, Confirmation of Charges Decision (Pre-Trial Chamber) 15 June 2009, para. 246. For fuller details, see https://www.icc-cpi.int/car/bemba, accessed 29 October 2018.

instead of pronouncing a NIAC.[18] Unhelpfully, therefore, such conflicts are only 'categorised' in the course of litigation and, temporally, often well after the event.[19] Critically, the status of captured persons will depend on the characterisation of the conflict. The category of POW only exists in IAC (although parties can and infrequently do agree to apply the rules in Geneva Convention III). The captured insurgent within a NIAC would be classified as a criminal detainee or possibly a security internee. S/he could not be classified as a POW because this category does not exist as a matter of law in NIAC.

Why Have Different Rules and What are the Consequences?

It has been argued that IAC rules should simply be used in NIAC. Indeed, in a small number of conflicts, belligerents have elected to treat all detainees as POW.[20] This approach has a certain albeit superficial appeal.[21] For one thing, the rules for POWs are extensive, clear and uncontroversial. Furthermore, fighters may be incentivised to apply the rules of IHL if they know they will not be prosecuted upon capture. There is no evidential requirement, save for hearings regarding the determination of POW status if there is any doubt.[22] In the example above, the capturing force could retain the fighter as a POW 'until the cessation of active hostilities',[23] thus removing any compulsion to effect a rapid transfer to host state law enforcement authorities. Detained fighters could also be exploited for intelligence over a long period of time and there is no risk of breaching policy and law as a result of the requirement to transfer them to an assisted state suspected of possible rights abuses.

---

18 Sandesh Sivakumaran, 'IHL Challenges series: Typology of Conflicts Part IV: Who decides?' *ICRC Blog*, 28 March 2013, available at http://intercrossblog.icrc.org/blog/ihl-challenges-series-typology-of-conflicts-part-iv-who-decides, accessed 29 October 2018.

19 The latest ICTY exposition on the definition of NIAC can be found in the cases of *Prosecutor v Dusko Tadić (jurisdiction of the Tribunal)* ICTY case no. IT-94-1-AR72 (2 Oct 1995) – more details available at http://www.icty.org/case/tadic/4 and *Prosecutor v Ramush Haradinaj* ICTY case no. IT-04-84-T (3 April 2008), available at http://www.icty.org/x/cases/haradinaj/tjug/en/080403.pdf, accessed 29 October 2018.

20 This happened, for instance, in the 1991 Croatian conflict. See Sivakumaran, *The Law of Non-International Armed Conflict*, 302–05.

21 See Emily Crawford 'Unequal before the law: The case for the elimination of the distinction between international and non-international armed conflicts', *Leiden Journal of International Law* 441 (2007).

22 For details, see *Geneva Convention relative to the treatment of prisoners of war on 12 August 1949* Art. 5, available at http://www.un.org/en/genocideprevention/documents/atrocity-crimes/Doc.32_GC-III-EN.pdf, accessed 29 October 2018.

23 *Geneva Convention relative to the treatment of prisoners of war on 12 August 1949*, Art. 118.

However, the most important distinction between detainees and POWs is that POWs are not prosecuted for lawful acts of war, as they are protected by their 'combatant immunity', whereas detainee/insurgents are liable to prosecution.[24] States are extremely reluctant to grant POW status to detainees, fearing this could legitimise the criminal groups they fight or even encourage insurgencies.[25] Moreover, POWs are released following cessation of active hostilities. The conflicts in Afghanistan and Iraq have lasted over a decade – and the so-called GWOT arguably continues *ad infinitum*. Although amnesties are possible, they are not guaranteed.[26] In other words, a de facto POW could face the worst of both worlds – prolonged detention *and* a criminal trial. Perhaps predictably, attempts made in 1949 and 1974 to introduce combatant immunity to those involved in a NIAC failed,[27] unable to overcome the concerns highlighted above. These remain extant.[28] In short, therefore, the unsatisfactory contemporary reality is a dearth of rules regarding NIAC in IHL.

Legal Bases for Assisting States to Detain in NIAC

To better understand what regime or modalities affect detainees in the example above, it is necessary to examine the legal basis (or bases) of detention. As noted above, the right to detain in IAC is explicit and uncontroversial.[29] Conversely, numerous challenges have been made regarding the detention by assisting states in NIAC.[30] A number of bases for detention by assisting states in NIAC are possible. These include a right under IHL, via UNSCRs or under customary international law (CIL). In addition, while detention under host state law and sending state law is

---

24 Direct participation in hostilities by civilians is not listed as a grave breach of the Geneva Conventions or a war crime in the International Criminal Court's 1998 Rome Statute for either IAC or NIAC.

25 Akande, 'Classification of Armed Conflicts: Relevant Legal Concepts', 38.

26 For details, see *Protocol Additional to the Geneva Conventions of 12 August 1949, and relating to the Protection of Victims of Non-International Armed Conflicts (Protocol II) on 8 June 1977* Art 6, available at https://ihl-databases.icrc.org/ihl/INTRO/475?OpenDocument, accessed 29 October 2018.

27 Sivakumaran, *The Law of Non-International Armed Conflict*, 42–53.

28 See Waldemar Solf, 'Problems with the application of norms governing interstate armed conflict to non-international armed conflict', *Georgia Journal of International and Comparative Law* 13 (1983), 291–92; Wilmshurst agrees that the greatest difference between IAC and NIAC is the issue of combatant immunity. Wilmshurst, *International Law and the Classification of Conflicts*, 569.

29 This right is contained within Geneva Conventions III and IV, which apply to International Armed Conflicts.

30 See Gary Rowe, 'Is there a right to detain civilians by foreign armed forces during a non-international armed conflict?', *International and Comparative Law Quarterly* 61:3 (2012), 702.

also possible, these are deemed unsatisfactory for the reasons highlighted previously. Indeed, until the UK's Supreme Court case of *Serdar Mohammed*,[31] the basis for detention in NIAC was an unsettled and unsatisfactory legal area. Controversy still remains, in part because the UK government does not necessarily accept the judgement in its entirety. With this in mind, each possible basis is examined below.

A Right under International Humanitarian Law

There is no *explicit* authority to detain in NIAC within the Geneva Conventions or within its two Additional Protocols (APs).[32] CA3 refers to 'parties to the conflict' and later to 'armed forces'. The drafting history seems to assume that the authority to detain lies solely with the state in which conflict occurs, the logic being that the assisting states occupy a secondary, supporting role, acting at the behest of the primary actor – the assisted state. APII similarly recognises administrative detention in Articles 5 and 6. APII refers to a conflict between '*its* armed forces' (emphasis added) and 'organised groups'. A literal interpretation suggests the right to detain only applies to the state's armed forces and opposition. A purposive interpretation would see assisting states equally covered, not least because the state has asked them to assist.

A powerful argument in support of IHL authorisation is that, if a party has the authority to kill in NIAC, surely it must also have the authority to detain. Otherwise, the negative corollary is that it would be lawful to kill, but not to capture – a position that appears contrary to the fundamental principles of IHL.[33] However, these arguments were unequivocally rejected in *Serdar Mohammed*, notably by Lord Sumption, who stated explicitly that CA3 'does not in terms confer a right of detention'.[34] In short, IHL regulates, but does not authorise, detention. If such a ruling is accepted (a position opposed by many), then the right to detain in international law must therefore derive from CIL or the authority of the UNSC.

---

31   For details of *Al-Waheed and Serdar Mohammed v Ministry of Defence* [2017] UKSC 2 (*Serdar Mohammed*), see https://www.supremecourt.uk/cases/docs/uksc-2014-0219-judgment. pdf, accessed 29 October 2018.

32   The Additional Protocols of 1977 acted as further amplification of the Geneva Conventions and codified Customary International Law further. Additional Protocol I dealt with rules under IAC, whilst Additional Protocol II dealt with rules under NIAC. The applicability is subject to a state's signature and ratification.

33   Ryan Goodman, 'The Detention of Civilians in Armed Conflict', *American Journal of International Law* 103:1 (2009), 55–56; Marco Sassóli, 'The International Legal Framework for Stability Operations: When May International Forces Attack or Detain Someone in Afghanistan?', *International Law Study Series*, 85 (2009), 454.

34   See *Serdar Mohammed*, para. 12.

A Right under Customary International Law

For a CIL rule to exist, two elements are required: (virtually) uniform state practice of the rule and a rule that is followed on the basis that it is required as a matter of law (*opinio juris*).[35] Without a doubt, there is much state practice of detention in NIAC. The major issue, however, is whether or not it is undertaken 'as required by law'. The UK has argued that the recent Copenhagen Process was evidence of *opinio juris* as to the right for assisting states to detain in NIAC. This involved 24 states (including the permanent five members of the UNSC), observers from the AU, the EU, NATO and the ICRC. Delegates agreed 'that detention is a necessary, *lawful* and *legitimate* means of achieving the objectives of military operations' (emphasis added).[36] Critically, while the ICRC also agreed that states have a right to detain in 'all forms of armed conflict', no consensus was reached concerning the relevant modalities.

Furthermore, it is highly debatable that 24 states constitute the degree of universality of state practice required for custom. The Copenhagen study also explicitly stated that it was not intended to create legal obligations or authorisations under international law.[37] Moreover, the right was not recognised by the respected International Law Commission's Customary Law Study, nor was it considered at the Experts' Meeting on Detention in NIAC. Finally, in *Serdar Mohammed*, the notion that the right was currently CIL was rejected, as the two elements noted above had not yet been fully met, although, tantalisingly, Lord Sumption conceded that it could be met at some future date.[38] Therefore, in the eyes of the UK Supreme Court at least, there is little judicial support to detain in a NIAC under custom.

---

35 The subjective belief is that a norm is legally binding, which, together with general state practice, equates to customary international law. See, for example, the *North Sea Continental Shelf Cases* [1969] ICJ Rep 3 – available at http://opil.ouplaw.com/view/10.1093/law:icgj/150icj69.case.1/law-icgj-150icj69 accessed 29 October 2018.

36 See Principle 3 of The Copenhagen Process. Danish Ministry of Defence, *The Copenhagen Process on the Handling of Detainees in Military Operations*.

37 Danish Ministry of Defence, *The Copenhagen Process on the Handling of Detainees in Military Operations*, 1, para. II. For a brief critique of the Copenhagen Process see Jacques Hartmann, 'The Copenhagen Process: Principles and Guidelines', *EJIL: Talk!* November 2012, available at https://www.ejiltalk.org/the-copenhagen-process-principles-and-guidelines/, accessed 29 October 2018; Jonathan Horowitz, 'Interlocutory note to the Copenhagen Process, Principles and Guidelines on the handling of detainees in military operations', *ASIL International Legal Materials* 51:6 (2012), 1364–66, available at https://papers.ssrn.com/sol3/papers.cfm?abstract_id=2228372, accessed 29 October 2018.

38 See *Serdar Mohammed*, para. 16.

A Right under a UN Security Council Resolution

Absent authority under IHL, CIL or host/sending state law, one possible lawful basis for detention could be found in a UNSCR. Under Chapter VII, Article 42 of the UN Charter, the UNSC can authorise an almost limitless range of actions, binding on all states. Such authorisations might include an authorisation to detain.[39] However, debate surrounds whether the commonly used formula of 'all necessary measures' is sufficient to imply authority to detain. Despite widespread use of detention in UN operations (including Kosovo, Afghanistan and Iraq), only one UNSCR, concerning Iraq, has mentioned detention specifically.[40] Thus it is difficult to maintain that a precedent had been set.

In the *Maya Evans* case, the UNSCR concerned Afghanistan, where detention was not explicitly mentioned. The UK court acknowledged 'a vital element of fulfilling the UN mission is to capture persons who threaten the security of Afghanistan' alluding to the authority under the UNSCR, apparently recognising that detention was so authorised[41] but only for short periods of time.[42]

Critically, a putative authorisation found in 'all necessary measures' gives no guidance as to the duration or grounds for detention. Human rights bodies have yet to pronounce on whether the UNSCR constitutes a 'lawful authority' as required under the 1996 International Covenant on Civil and Political Rights (ICCPR) and the 1950 European Convention on Human Rights (ECHR).[43] The Chatham House and ICRC expert meeting on detention concluded that, whilst there was agreement that a UNSCR

---

39 International Committee of the Red Cross, *Expert meeting on procedural safeguards for security detention in non-international armed conflict*, 869.

40 United Nations, *United Nations Security Council Resolution 1546 on the situation between Iraq and Kuwait on 8 June 2004* (New York: United Nations, 2004), available at http://unscr.com/en/resolutions/1546, accessed 28 October 2018.

41 Detention in Iraq was explicitly authorised under UNSCR 1546 (2004) – details above. UNSCR 1386 regarding Afghanistan makes no reference to detention but was 'taken to mean' authorisation of ISAF to use 'all necessary measures to fulfil its mandate', which was interpreted as implied authority only to detain temporarily. United Nations, *United Nations Security Council Resolution 1386 on the situation in Afghanistan on 20 December 2001* (New York: United Nations, 2001) available at http://unscr.com/en/resolutions/1386 accessed 29 October 2018. However, the Court in the Maya Evans case referred to legal advice that confirmed there existed 'no legal basis upon which the UK could intern such individuals, [and they] must be transferred to the Afghan authorities'. For details of this case, see http://www.internationalcrimesdatabase.org/Case/1194, accessed 29 October 2018.

42 Rowe, 'Is there a right to detain civilians by foreign armed forces during a non-international armed conflict?', 706.

43 Rowe, 'Is there a right to detain civilians by foreign armed forces during a non-international armed conflict?', 706.

could provide a legal basis, there was no agreement as to the specificity of language required in the resolution.[44]

In the UK the Supreme Court in *Serdar Mohammed* concluded that a UNSCR could indeed act as the legal basis for detention in such operations, citing European Court of Human Rights (ECtHR) jurisprudence regarding Kosovo in recognition that the authority to detain could also be implied.[45] Clearly, there is no UNSCR 'silver bullet'. Whether detention is 'necessary' for the purposes of the UNSCR 'depends primarily on the specific mandate, on the general context and on conditions and initiations laid down'.[46] Thus, a thorough reading of both the mandate itself and circumstances are required.

A UNSCR can be hugely helpful to assisting states in NIAC, even if different national interpretations frustrate agreement about the precise parameters of a mission.[47] Conversely, the absence of a UNSCR – a stark reality in contemporary international politics – leaves states without even this possible source of authorisation. If a legal basis has been identified for detention in a NIAC, other questions remain: what do the rules say regarding how long detainees can be detained? What rules govern their right (if any) to challenge their detention? Can they represent themselves and provide evidence? What are the rules regarding transfer? UNSCRs are drafted in many cases with intended ambiguity. Absent such specific detail, a brief discussion of these modalities is necessary.

The Modalities of Detention in NIAC

If detention is deemed to have been authorised under a UNSCR, such authorisations will rarely (if ever) set out the modalities noted above. UNSCRs are, in essence, political compromises, purposefully drafted in wide terms to accommodate the differing views and interests of states. NIAC operations share characteristics of IAC regarding the scale and intensity of military operations, but concurrently, because forces acting against the government have no legal right to take part in hostilities, such operations also take on a law enforcement paradigm. The question is – regarding detention modalities – which rules take precedence: IHRL, IHL or domestic law? Failure to

---

44 Rowe, 'Is there a right to detain civilians by foreign armed forces during a non-international armed conflict?', 869.

45 See *Serdar Mohammed*, paras 18–30.

46 *Serdar Mohammed*, para. 26.

47 The controversy over the legality of the invasion of Iraq in 2003 following numerous UNSCRs is an obvious example of this.

identify and apply the correct modalities presents the assisting state with greater legal risks. Where might such rules exist?

As observed earlier, the substantive rules in IHL governing NIAC are sparse compared to those within IAC. IAC has an entire treaty (Geneva Convention III) dedicated to the treatment and rules concerning POWs. Much of Geneva Convention IV outlines in detail the detention/internment of civilian detainees. Due to the lack of state consensus, none of the advances made within the last 20 years regarding the applicable law in NIAC address the lack of 'hard rules' within NIAC detention. The ICRC has made recommendations,[48] but this is not 'hard law' (substantive law), nor has it acquired universal acceptance. The Copenhagen Process also attempted to set down detention rules within NIAC and resolve some of these issues.[49] However, as noted above, the principles agreed specifically do not create legal obligations or authorisations.[50]

Therefore, the substantive law within IHL regarding NIAC detainees remains in a single article of the Geneva Conventions (CA3) and within the three articles of APII. This sets out, inter alia, prohibited acts such as 'violence to life and person, in particular murder of all kinds, mutilation, cruel treatment and torture; the taking of hostages and outrages upon personal dignity, in particular humiliating and degrading treatment'. There are no details concerning grounds, review or representation.

The relevant articles of APII (4, 5 and 6) are, again, general in nature. Article 6 APII affords some detail, in that basic fair trial rights are afforded to detainees, which mirror the fair trial rights under the ICCPR, including the requirement to be informed 'without delay' of the particulars of *any offence* and 'all necessary rights and means of defence'; the presumption of innocence, the right to silence and trial in his presence; and the right not to be compelled to testify against himself. However, there is no detail regarding the meaning of 'without delay' and 'all necessary means and rights to defense'. Such provisions refer to the trial of the substantive offence (if any charge is brought), not the power to detain before the criminal process has begun. The rules are scant because, throughout the negotiating process of APII, familiar arguments regarding state sovereignty and non-intervention abounded, leading to

---

48 Jelena Pejić, 'Procedural Principles and Safeguards for Internment/Administrative Detention in Armed Conflict and Other Situations of Violence', *International Review of the Red Cross* 87:858 (2005), 375–91.

49 See Thomas Winkler, 'The Copenhagen Process on Detainees: A Necessity', *Nordic Journal of International Law* 78 (2010), 489–98.

50 Danish Ministry of Defence, *The Copenhagen Process on the Handling of Detainees in Military Operations*, paras II, 25 and para. 16.2.

considerable compromises.[51] Critically, it was also presumed there would be a functioning human rights-compliant legal process to hand such persons over to.

The presumption that the host state's legal system would cater for detention is an aspiration not reflected in contemporary practice. NIAC predominately occurs within fragile, failing or failed states, where procedural safeguards either do not exist or the lack of adequate legislation, infrastructure and resources make them meaningless. Applying host national systems in many cases simply does not work. Host state law may conflict with assisting wider human rights obligations or the host may be unable or unwilling to abide by its own obligations.[52]

The laws of the host state may not contain provisions necessitating the giving of reasons for detention or any established procedure whereby the lawfulness of detention can be challenged. Providing reasons to those individuals who have been deprived of their liberty can be difficult; detention often occurs via sensitive intelligence, such as the protection of informants, the identity of whom is clearly sensitive, leaving the detainee unable to challenge the source. For these reasons, detention is frequently challenged not in the host state but via the courts of the assisting states through *habeas corpus* applications, judicial review or human rights litigation. Once again, the availability of such remedies is divergent.

Given the lack of substantive law in IHL, CIL takes on heightened importance. Of the 161 Customary Rules identified in the ICRC customary law study, 148 were seen to apply in NIAC as well as IAC. However, CIL still has its defects, such as prohibiting 'arbitrary deprivation of liberty' without defining 'arbitrary'.[53] State practice appears to support the ICRC stance that detention should occur 'exceptionally' and only for 'imperative reasons of security' (a phrase which will assume great importance later in this chapter). This is a high threshold, emanating from the civilian internment standard in IAC and within Geneva Convention IV.[54] This could include evidence of the detainee having taken a direct part in hostilities, with the attendant difficulties attached to ascertaining

---

51 The draft of Additional Protocol II was reduced from 48 articles to 28. See Adam Roberts and Richard Guelff (eds), *Documents on the Laws of War* (Oxford: Oxford University Press, 2001), 481–82.

52 Customary Law Study (n 7), 304, referring to the UN Special Rapporteur's report to the UNCHR in 1987 regarding Afghan authorities 'acting without any respect for the international human rights obligations for which they have assumed'.

53 See *Rule 99, which notes that 'Arbitrary deprivation of liberty is prohibited'. For details, see* Doswald-Beck and Henckaerts (eds), *International Humanitarian Law Volume I: Rules.*

54 For details, see *Geneva Convention relative to the protection of civilian persons in time of war of 12 August 1949*, Arts 41–43 and 68–78, available at http://www.un.org/en/

what constitutes such participation.[55] It does, however, seem to meet the needs of the security of the state and the rights of the individual, although it applies in the strict sense to IAC and not to NIAC.

General International Law

Other international law – such as the 1984 United Nations Convention Against Torture (UNCAT) and the 1951 Convention Relating to the Status of Refugees (hereafter the Refugee Convention) – will apply in NIAC, but only to the contracting parties and subject to ratification and, in the case of the UK, on an article-by-article basis. It is this area of law, in addition to applicable human rights law, which prevents a state forcibly transferring a person from its jurisdiction where there is a real risk of death or torture (known as *non-refoulement*). Thus, assisting forces cannot simply transfer detainees where such a risk exists. The UK goes further, stating that its forces will not transfer any captured person (including to the host state) where there is 'a real risk at the time of transfer that the captured person (CPERS) will suffer torture, serious mistreatment or be subjected to unlawful rendition'.[56] A pragmatic solution is to obtain credible assurances regarding post transfer treatment, yet these are not without controversy and must be borne out in practice.[57] As the UN General Assembly has recently stated, 'diplomatic assurances ... should not be used as a loophole to undermine the principle *of non-refoulement*'.[58]

International Human Rights Law

In a NIAC, detention is governed by IHL and the domestic law of that state, which could include IHRL. Controversy surrounds the extent to which one regime 'trumps the other' (discussed further below) and

---

genocideprevention/documents/atrocity-crimes/Doc.33_GC-IV-EN.pdf, accessed 28 October 2018.

55 Pejić, 'Procedural Principles and Safeguards for Internment/Administrative Detention in Armed Conflict and Other Situations of Violence', 95.

56 Development, Concepts and Doctrine Centre, *Joint Doctrine Publication 1-10 (Captured Persons)* para. 1205.

57 See Başak Çali and Stewart Cunningham, 'Part 1: A few steps forward, a few steps sideways and a few steps backwards: The CAT's revised and updated GC on Non-Refoulement', *EJIL: Talk!* March 2018, available at https://www.ejiltalk.org/part-1-a-few-steps-forward-a-few-steps-sideways-and-a-few-steps-backwards-the-cats-revised-and-updated-gc-on-non-refoulement/, accessed 29 October 2018.

58 United Nations Committee against Torture, *General Comment No. 1 (2017) on the Implementation of Article 3 of the Convention in the Context of Article 22* (New York: United Nations, February 2017), 20, available at https://www.ohchr.org/Documents/HRBodies/CAT/GCArticle3/CAT-C-GC-1.pdf, accessed 29 October 2018.

whether non-criminal detention is lawful without derogation from the universal human rights system, namely the ICCPR, even if judicial review is provided for.[59] The UK Supreme Court has ruled that the right to judicially challenge detention via *habeas corpus* still exists, even in times of conflict.[60] Assisting states often obtain immunity from host state law by way of a Status of Forces Agreement (SOFA) or a similar instrument.[61] However, these agreements only govern jurisdiction over the assisting state's forces, not over the detainees held by them. Bilateral agreements covering detention between hosts and assisting states are common, but problematic. Such agreements invariably contain assurances regarding treatment and transfers, but cannot always be relied upon. Moreover, the ECtHR has ruled that such arrangements cannot override assisting states' human rights obligations,[62] nor can they absolve the state where it seeks to rely on the host state authority to detain and thereby avoid ECHR liability.[63] Such responsibilities have the potential to severely curtail ECHR states' ability to conduct detention within operations in states with poor human rights records or capabilities.[64] The lack of substantive IHL regarding detention modalities in NIAC is a critical weakness of IHL. Unsurprisingly, courts have tended to use IHRL to fill the vacuum,

---

59   Pejić, 'Procedural Principles and Safeguards for Internment/Administrative Detention in Armed Conflict and Other Situations of Violence', 90.

60   See *Serdar Mohammed*, para. 100; The Human Rights Committee, General Comment No. 29, para. 16 (New York: United Nations, 2001), available at http://hrlibrary.umn.edu/gencomm/hrc29.html, accessed 28 October 2018.

61   See, for example, Coalition Provisional Authority, *Order no 17 (Revised) Status of the Coalition Provisional Authority, MNF-Iraq, Certain Missions and Personnel in Iraq* (Washington DC: Coalition Provisional Authority, 2004) paras 2(1), 2(3), available at http://www.refworld.org/docid/49997ada3.html, accessed 29 October 2018; United Nations, 'Military Technical Agreement between ISAF and the Interim Administration of Afghanistan' (New York: United Nations, January 2002) Annex A, para. 3, available at https://reliefweb.int/report/afghanistan/afghanistan-militarytechnical-agreement-between-international-security, accessed 29 October 2018. From a UN perspective, *The Convention on Privileges and Immunities of the United Nations 1946*, art VI, para. 1 applies *mutatis mutandis* to ISAF forces, para. 1. Regarding criminal liability for acts within a host state see United Nations, *Criminal Accountability of United Nations Officials and Experts on Mission* (New York: United Nations, September 2007) para. 63, available at http://legal.un.org/docs/?symbol=A/62/329, accessed 28 October 2018.

62   European Court of Human Rights, *Al-Jedda v The United Kingdom App. No. 27021/08* (Strasbourg: ECtHR, 7 July 2011), available at http://www.bailii.org/eu/cases/ECHR/2011/1092.html, accessed 29 October 2018.

63   European Court of Human Rights, *Al-Sadoon and Mufdhi v The United Kingdom App. No. 61498/08* (Strasbourg: ECtHR, March 2010), available at http://www.asylumlawdatabase.eu/en/content/ecthr-%E2%80%93-al-saadoon-and-mufdhi-v-united-kingdom-application-no-6149808-30-june-2009-%E2%80%93, accessed 29 October 2018.

64   Pejić, 'Procedural Principles and Safeguards for Internment/Administrative Detention in Armed Conflict and Other Situations of Violence', 92.

making the importance of the very prescriptive human rights regime difficult to understate.

Human Rights Applicability on the Battlefield

The American and European Conventions on Human Rights require each party to ensure rights to everyone 'within their jurisdiction' – a concept that the ECtHR has interpreted broadly,[65] including detention centres during NIAC.[66] The ICCPR requires 'protection of the rights and freedoms of those within its territory *and* jurisdiction' (emphasis added). The US views this criterion as being cumulative, that is an act occurring within its geographical area and legal jurisdiction.[67] This is a vocal, but minority view, contrary to the opinions of the International Court of Justice (ICJ),[68] the UN Human Rights Committee[69] and many other states.[70]

---

65 European Court of Human Rights, *Loizidou v Turkey, App. No. 15318/89* (Strasbourg: ECtHR, December 1996) available at http://www.asylumlawdatabase.eu/en/content/ecthr-loizidou-v-turkey-application-no-1531889-18-december-1996, accessed 29 October 2018; European Court of Human Rights, *Cyprus v Turkey, App. No. 25781/94* (Strasbourg: ECtHR, May 2001), available at http://www.asylumlawdatabase.eu/en/content/ecthr-cyprus-v-turkey-application-no-2578194-10-may-2001, accessed 28 October 2018; European Court of Human Rights, *Medvedyev and Others v France, App. No. 3394/03* (Strasbourg: ECtHR, March 2010), available at file:///C:/Users/User/Downloads/002-1015%20(1).pdf, accessed 28 October 2018.

66 European Court of Human Rights, *Al-Jedda v The United Kingdom App. No. 27021/08* (Strasbourg: ECtHR, 7 July 2011), available at http://www.bailii.org/eu/cases/ECHR/2011/1092.html, accessed 29 October 2018.

67 For the US argument see Department of State, *United States Second and Third Periodic Report of the United States of America to the UN Committee on Human Rights Concerning the International Covenant on Civil and Political Rights* (Washington DC: State Department, October 2005), available at https://www.state.gov/j/drl/rls/55504.htm, accessed 29 October 2018.

68 For details on the relevant International Court of Justice cases, see the *Wall case*, paras 107–12, available at http://www.unrod.org/docs/ICJ-Advisory2004.pdf, accessed 29 Ocber 2018 and *DRC v Uganda* paras 216–17, available at https://www.asil.org/insights/volume/10/issue/1/case-concerning-armed-activities-territory-congo-icj-finds-uganda-acted, accessed 29 October 2018.

69 United Nations Human Rights Committee, *General Comment 31: The Nature of the General Legal Obligation Imposed on States Parties to the Covenant* (Geneva: United Nations, May 2004) para. 10, available at https://www.ohchr.org/EN/Issues/Education/Training/Compilation/Pages/c)GeneralCommentNo31TheNatureoftheGeneralLegalObligation-ImposedonStatesPartiestotheCovenant(2004).aspx, accessed 29 October 2018.

70 See, for instance, UK Ministry of Defence, *The Joint Service Manual of the Law of Armed Conflict JSP 383* (London: Ministry of Defence, 2004) para. 11.19, available at https://assets.publishing.service.gov.uk/government/uploads/system/uploads/attachment_data/file/27874/JSP3832004Edition.pdf, accessed 29 October 2018.

Whilst *Banković*[71] previously ruled that ECHR jurisdiction rights could not be 'divided and tailored' and must be applied in whole or not at all, this rationale has been successively eroded. The UK has conceded that jurisdiction arises where there is exclusive control over an individual (i.e. in detention),[72] while the ECtHR cases of *Al-Skeini* and *Al-Jedda*[73] established that jurisdiction was found in cases where effective control is proven, either through 'total control' over the individual (where the relevant human rights will be afforded) or 'effective control' over a territory (where all human rights requirements will be afforded). In either case, detention fulfils the criteria for human rights to apply. It is right to note that both of these cases concerned actions during NIAC – where the UK had argued that IHRL did not apply and that IHL was the proper legal regime. They thus represented a significant shift in the IHL versus IHRL debate.

States have also argued that IHRL does not apply in conflicts in other circumstances. They have also attributed responsibility for their actions to other bodies, such as the UN (which is not a party to any IHRL treaty). This was accepted in *Behrami and Saramati*, although subsequent reliance upon this argument has failed.[74] A highly specific authorisation would be required giving specific control to the UN as to specific acts and detailed provisions regarding modalities of detention that must exist. In practice, states retain national control over detention and authorisation but the UN remains distinct from actual direction and control and therefore attribution.[75] While the exception in *Behrami* survives, its application is regarded as limited due to the fact that it concerned a very fact-specific case, where the UN had direct control over detention and the operation via

---

71   European Court of Human Rights, *Banković and Others v Belgium and Others*, App. No. *52207/99* (Strasbourg: ECtHR, December 2001), available at https://hudoc.echr.coe.int/eng#{%22itemid%22:[%22001-22099%22]}, accessed 29 October 2018.

72   In the ECtHR case *Hess v United Kingdom* (1975) the UK did not have exclusive control. For details, see file:///C:/Users/User/Downloads/001-73854.pdf, accessed 29 October 2018.

73   European Court of Human Rights, *Al-Skeini and Others v The United Kingdom*, App. No. *55721/07* (Strasbourg: ECtHR, July 2011), available at http://www.asylumlawdatabase.eu/en/content/ecthr-%E2%80%93-al-skeini-and-others-v-united-kingdom-application-no-5572107-7-july-2011, accessed 29 October 2018.

74   European Court of Human Rights, *Al-Jedda, Behrami and Saramati v France and Others* (*Behrami*) *App. No 71412/01 and 78166/01* (Strasbourg: ECtHR, May 2007), available at https://www.tjsl.edu/slomansonb/3.1_UN_attrib.pdf, accessed 29 October 2018. For a scathing critique of the ECtHR's reasoning in *Behrami*, see Marko Milanović and Tatjana Papić, 'As bad as it gets: The European Court of Human Right's Behrami and Saramati Decision and General International Law', *International and Comparative Law Quarterly* 58 (2009), 267–96.

75   See the International Law Commission, *Report on the Draft Articles on the Responsibilities of International Organisations Report 58th Session* (New York: United Nations, 2006), available at http://legal.un.org/docs/?path=../ilc/reports/2006/english/chp7.pdf&lang=EFS-RAC, accessed 29 October 2018.

a clear chain of command. Therefore, in many instances, IHRL will apply to detention. It can be safely concluded that, with regard to detention in NIAC, human rights law will indeed apply. The difficult question that remains is to what extent.

The Applicability of Human Rights in NIAC

As Pejić rightly asserts 'it is neither helpful or sufficient to state that human rights law continues to apply in armed conflict without elaborating on what that means in practice'.[76] The view that IHL applies in armed conflict and IHRL applies in situations outwith armed conflict is now moribund. However, as the ICJ (in)famously stated:

> [S]ome rights may be exclusively matters of international humanitarian law; others may be exclusively matters of human rights law; yet others may be matters of both these branches of international law.[77]

The difficulty with applying human rights law in full in NIAC is that the grounds to detain in human rights law are exhaustive. Article 5 of the ECHR (the right to liberty) authorises detention only under 'a procedure *prescribed by law*' (emphasis added). It states:

> Everyone has the right to liberty and security of person. No one shall be deprived of his liberty save in the following cases and in accordance with a procedure prescribed by law:
>
> (a) the lawful detention of a person after conviction by a competent court;
> (b) the lawful arrest or detention of a person for non-compliance with the lawful order of a court or in order to secure the fulfilment of any obligation prescribed by law;
> (c) the lawful arrest or detention of a person effected for the purpose of bringing him before the competent legal authority on reasonable suspicion of having committed an offence or when it is reasonably considered necessary to prevent his committing an offence or fleeing after having done so;
> (d) the detention of a minor by lawful order for the purpose of educational supervision or his lawful detention for the purpose of bringing him before the competent legal authority;

---

76 International Law Commission, *Report on the Draft Articles on the Responsibilities of International Organisations Report 58th Session* (New York: United Nations, 2006).

77 International Court of Justice, *Legal Consequences of the Construction of a Wall in the Occupied Palestinian Territories* (The Hague: ICJ, 2004), 136, available at http://www.unrod.org/docs/ICJ-Advisory2004.pdf accessed 29 October 2018.

(e)   the lawful detention of persons for the prevention of the spreading of infectious diseases, of persons of unsound mind, alcoholics or drug addicts or vagrants;

(f)   the lawful arrest or detention of a person to prevent his effecting an unauthorised entry into the country or of a person against whom action is being taken with a view to deportation or extradition.

Furthermore, anyone detained must 'be informed promptly' of the reasons for his arrest and any charge. Detainees 'shall be brought promptly before a judge or other officer authorised by law to exercise judicial power and shall be entitled to trial within a reasonable time or to release pending trial'. Detainees are also entitled to challenge the lawfulness of the detention 'speedily by a court'. Similar, but less restrictive, criteria are contained within Article 9 of the ICCPR.

If unmodified, these are extremely onerous requirements on the battlefield and, in many cases, cannot be satisfied. Critically, none of the grounds listed above allow detention for force protection or mission accomplishment – grounds frequently used by armed forces on operations in NIAC for obvious and sensible reasons. Such operations have been described as 'far too complex and brutal a phenomenon to be capable of being constrained by rules designed for peacetime'.[78] Restrictive regulations may be apposite in Strasbourg but do not fit neatly into the complex fragile states and 'wars amongst the people' that feature in the contemporary security environment.[79]

Derogation

It is possible, in some circumstances, to 'opt out' of human rights obligations. However, the ICCPR only allows derogation from specific obligations '[i]n time of war or other public emergency threatening the life of the nation'.[80] This implies that the treaty applies in armed conflict

---

78   Christopher Greenwood, 'Rights at the Frontier: Protecting the Individual in Time of War' in Barry Rider (ed.) *Law at the Centre* (Amsterdam: Kluwer Law International 1999), 293.

79   In 2011 it took the author three days to travel from southern Helmand to Kandahar (Sarposa) prison. This involved a fixed wing flight, a helicopter flight (which was delayed by 24 hrs by a sand storm) and a road move using five heavily armoured vehicles through Kandahar city, known to be targeted by insurgents.

80   See United Nations, *International Covenant on Civil and Political Rights* (New York: United Nations, 1966) Article 4, available at http://www.un-documents.net/iccpr.htm, accessed 29 October 2018; Council of Europe, *European Convention on Human Rights* (Strasbourg: Council of Europe, 1950) Article 15, available at https://www.echr.coe.int/Documents/Convention_ENG.pdf, accessed 29 October 2018.

unless a state derogates and only for certain rights;[81] critically, some rights, such as the right to life and the right not to be tortured, are non-derogable.[82] The UK Supreme Court confirmed that derogation from the ECHR is possible only:

> [I]n times of war or other public emergency threatening the life of the nation *seeking to derogate* (emphasis added), and only then to the extent strictly necessary required by the exigencies of the situation ... It is hard to think that these conditions could ever be met when a state had chosen to conduct an overseas peacekeeping operation, however dangerous the conditions, from which it could withdraw.[83]

To date, no state has ever successfully derogated from IHRL obligations for acts committed outside its own borders during armed conflict, although the UK in 2016 stated it retains the right to do so if required.[84] Whilst derogation is technically possible, there are huge political and presentational difficulties with any state derogating from human rights provisions, particularly during a UN mission. Although outwith the scope of this chapter, there is also a debate as to whether derogation could, in any event, avoid the issues set out above, in part or at all.

Human Rights – But Modified

Relying on the Grand Chamber ECtHR case of *Hassan*, the UK Supreme Court's judgment in *Serdar Mohammed* offers some useful insights.[85] This case ruled that human rights requirements can be read to accommodate IHL and need not be read in isolation.[86] The Court held that the exhaustive

---

81 This approach was confirmed by the ICJ in *the Wall* case and was followed in its *DRC v Uganda* case and *Nuclear Weapons* Advisory Opinion.

82 See United Nations, *International Covenant on Civil and Political Rights* (New York: United Nations, 1966) Article 6; Council of Europe, *European Convention on Human Rights*, Article 15.

83 Supreme Court, *Smith (No 2) et al. v Ministry of Defence [2013] UKSC 41* (London: House of Lords, 2013) available at https://www.supremecourt.uk/decided-cases/docs/UKSC_2012_0249_Judgment.pdf, accessed 29 October 2018.

84 Pejić, 'Procedural Principles and Safeguards for Internment/Administrative Detention in Armed Conflict and Other Situations of Violence', 93.

85 European Court of Human Rights, *Hassan v The United Kingdom*, App. No. 29750/09 (Strasbourg: ECtHR, September 2014) available at file:///C:/Users/User/Downloads/002-10082.pdf, accessed 29 October 2018.

86 Put briefly, where a true conflict between laws regarding the same norm arises (i.e. between detention rules in IHL and IHRL), the more specific rule(s) will apply. See International Law Commission, *Fragmentation of International Law: Difficulties arising from the Diversification and Expansion of International Law: Report of the Study Group of the*

grounds in Art 5 ECHR should be read together with the Geneva Convention IV grounds – in other words, for 'Imperative Reasons of Security' (IRoS). Whilst there is still no ground that allows for detention purely for other purposes (including solely for intelligence),[87] detention can continue if, and only for as long as, the grounds of IRoS are met.

In terms of detention in NIAC, this is a significant judgment and offers greater clarity regarding some of the modalities of detention, which were either absent in IHL or overly restrictive in IHRL. *Serdar Mohammed* was, in many ways, a 'mixed judgement' with both positive and less positive outcomes for the UK government. However, detaining states wishing to rely on the same must exercise caution. The *Hassan* case occurred during an IAC, not an NIAC. Lord Sumption's view was that there should be no reason why the principle ground in IAC (IRoS) should not apply in both types of conflict may well be challenged, although the ECtHR's subsequent rejection of Serdar Mohammed's appeal would appear to confirm the judgment as 'good law' at the UK Supreme Court level at least.

Other issues also remain. The UK detention regime in Afghanistan (the NIAC in which Serdar Mohammed was detained) failed to fulfil the requirements under both IHRL and within the Geneva Conventions. The fundamental purpose of Article 5 ECHR and Articles 43 and 78 of Geneva Convention IV is to protect against arbitrary detention.[88] The Court stated that, to avoid arbitrariness, there must be both a legal basis for detention and sufficient means to challenge the lawfulness of that detention. Any regime must be sufficiently fair and have independent competent bodies (but not necessarily a court). The last point is of real practical assistance to states. Having established the basis under IHRL, any detention review process must be independent of the chain of command, both in fact and law.[89] The independence of any review board in such conflicts is 'resource

---

International Law Commission (New York: United Nations, 2006), 56–122, available at http://legal.un.org/ilc/documentation/english/a_cn4_l682.pdf, accessed 29 October 2018.

87 European Court of Human Rights, *Sakik and Others v Turkey, App. No. 87/1996/706/898-903* (Strasbourg: ECtHR, November 1997) para. 44, available at file:///C:/Users/User/Downloads/001-58117.pdf, accessed 29 October 2018; European Court of Human Rights, Öcalan v Turkey, App. No. 46221/99 (Strasbourg: ECtHR, May 2005), para. 104, available at http://www.freedom-for-ocalan.com/english/hintergrund/fall/CASE%20OF%20OCALAN%20v%20TURKEY.pdf, accessed 29 October 2018; European Court of Human Rights, *Medvedyev and Others v France, App. No. 3394/03* (Strasbourg: ECtHR, March 2010) para. 126, available at file:///C:/Users/User/Downloads/002-1015%20(1).pdf, accessed 28 October 2018.

88 *Al-Waheed and Serdar Mohammed v Ministry of Defence* [2017] UKSC 2 (Serdar Mohammed), available at https://www.supremecourt.uk/cases/docs/uksc-2014-0219-judgment.pdf, accessed 29 October 2018.

89 Lord Sumption opined that the nearest to fully impartial members of the Detention Review Committee were identified as the Political Advisor and the Legal Advisor.

heavy'. Other difficulties arise. Of assistance to assisting states, *Serdar Mohammed* ruled that the UK's detention regime under a 'Standing Operating Instruction' (known as SOI J3-9), an internal military policy, did satisfy the requirements of 'legal certainty'.[90] Furthermore, the amount of time to be brought 'promptly' before 'an officer authorised by law' was left as a matter of fact, with the court recognising the special circumstances in each case – but four days (96hrs) is 'probably reasonable'.

The review of detention is, however, critical. In Afghanistan, the only realistic process was under the military SOI J3-9 system. Whilst many aspects were compliant, critically, the reviewing authority was the same authority that authorised detention and no appeal body post-review existed. It therefore lacked *apparent* (not actual) independence. Of paramount importance, the detainee had no involvement in the process. In his comments, Lord Sumption offered four characteristics of a reviewing body that would support the process as offering a fair review. These are of particular import to any detaining force in a NIAC. The requirements were that:

1. The review procedure is explained to the detainee;
2. The detainee must be given the 'gist' of the evidence against them;
3. They must be given contact with the outside world to obtain evidence, and
4. They must be allowed to make personal representations and preferably be legally represented.

Save for the first point, such requirements are extremely problematic in an operational context. If the evidence against a detainee is based substantively or purely on sensitive intelligence, how can such evidence be made available to him? This is magnified if the intelligence is from a highly vulnerable human source, or worse, from a coalition partner sensitive about sharing intelligence, where such a source could be compromised. This Kafkaesque situation taxes even UK courts, such as the Special Immigration Tribunals, let alone a hard-pressed operational staff in the middle of a conflict zone. Beyond the specialist branches of the military, soldiers are not trained investigators, statements can be difficult to obtain and the battlefield is no place to carry out full scenes of crime processes.

Giving detainees access to 'the outside world' is equally problematic. Security can be compromised by such access and there remains a question of

---

90 'A body of rules which is enforceable, sufficiently specific, and operates within a framework of law including public law'. See European Court of Human Rights, *Medvedyev and Others v France.*

how exactly such evidence is to be gathered, given the locality and situations in which many insurgents are detained. Whilst personal representation could be arranged, the notion of representation by lawyers from the assisted state is perhaps naïve. In southern Afghanistan, there were few, if any, defence counsel practising. One option is for Service lawyers to assist, but this risks offending the 'apparent bias' principle. Other options include the use of technology, such as video television conferencing, although this is highly dependent on a static and mature infrastructure being in place. On a more positive note, breaches of these procedural obligations (ECHR Art 5(4)) would not render the detention unlawful, although breaches do attract legal and reputational consequences.

Conclusion

The most common form of conflict today is not between states but is indeed 'war amongst the people'. As this chapter has shown, the law governing detention in such conflicts can be vague, incomplete and, where the blanket application of human rights is concerned, highly impractical. Nonetheless, detention remains critical for mission success. A clear-eyed assessment suggests that the challenges detailed above can be overcome. Detention operations should not be avoided simply because of the risks and complexity. Adopting this course has proved unrealistic in ground operations. Like all military activity, recognition and adaptation requires pre-emptive thought, doctrinal recognition, training and war-gaming by all stakeholders. Detention delivers a district military effect and must be thought of in the same vein.

Following cases from the ICJ, ECtHR, commentary from IHRL bodies and domestic courts (including the UK), the direction of travel indicates the increased applicability of IHRL, even in situations of high intensity internal armed conflict. IHRL bodies and national courts must allow IHL provisions to inform their decisions so that a true symbiotic relationship between the regimes can be achieved. Until a degree of state consensus can be achieved, and better nuanced discourse between the two IHL and IHRL camps occurs, ad hoc state and situation-specific 'solutions' will remain prevalent, as will further fragmentation and dissonance of the two legal regimes.[91]

In the wake of *Serdar Mohammed* some certainty (for the UK at least) can be found as to the legal basis for assisting states to detain in NIAC:

---

91 Daniel Bethlehem, 'The Relationship between International Humanitarian Law and Human Rights Law in situations of Armed Conflict', *Cambridge Journal of International and Comparative Law* 2:2 (2013), 195.

UNSCR can be applicable, but only where the context and language permits it. The situation where operations ensue without a UNSCR (as in Kosovo) remains unresolved. This is a very realistic scenario given the political dynamics at present. Of course, *Serdar Mohammed* is a UK case and, whilst highly influential, does not bind other states. Furthermore, the UK courts have tended to reach sensible, pragmatic solutions only for them to founder later at the ECtHR level. That said, *Hassan* (which *Serdar* followed, if not analogously) is a Grand Chamber ECtHR case and so should offer some protection for ECHR assisting states.

Real challenges still exist regarding the modalities of detention, but these are not insurmountable. The precise boundaries as to the grounds of 'imperative reasons of security' remain unclear and the drafting history suggests a high bar. However, sound doctrine based on the principles set out in cases such as *Serdar Mohammed* offer a welcome handrail.[92] Likewise, the independence of any review body will require resourcing and structural independence. The utilisation of Service lawyers may well be the key and deployed military judges may yet be seen. Calls for representation for the detainee seem eminently sensible, but this does not take into account the security concerns of high value targets communicating externally post-capture. Whether such resources, including competent defence legal representatives can be provided in the host state remains problematical.[93] Should the problems inherent within 'wars amongst the people' regarding detention pervade, the UN, and indeed the world, may struggle to find states willing to support failing states because, ironically, the current legal system prevents them from doing so without incurring significant legal risk.

---

92 Guidance as to what 'Imperative Reasons of Security' means is set out but is far from prescriptive. The 'Security of the State' was not defined in a 'more concrete fashion' and thus 'left very largely to Governments to decide the measure of activity prejudicial to the internal or external security of the State which justifies internment'. Subversive activity could include actions which are of direct assistance to an enemy that 'threaten security of the country' or where there is 'serious and legitimate reason to think that they are members of organisations whose object is to cause disturbances, or that they may seriously prejudice its security by other means, such as sabotage or espionage'. However, the mere fact that a person is a subject of an enemy Power cannot be considered sufficient. The commentary argues that there be 'good reason' to think that the person concerned, by his activities, knowledge or qualifications, represents a real threat to present or future security. The Convention stresses the exceptional character of measures of internment and only 'absolute necessity', based on the requirements of state security, can justify detention and 'only then if security cannot be safeguarded by other, less severe means'. For details, see United Nations, *Final Record of the Diplomatic Conference of Geneva of 1949 Vol III* (New York: United Nations, 1949), 126, available at https://www.loc.gov/rr/frd/Military_Law/pdf/Dipl-Conf-1949-Final_Vol-3.pdf, accessed 29 October 2018.

93 From personal experience, in 2011, less than a dozen properly trained defence counsels were known to exist in the whole of Afghanistan, let alone the highly dangerous region of Helmand and Kandahar, where the UK operated.

References

Amnesty International v Canada Case 2008 FC 336 (12 March 2008); *Hamdan v Rumsfeld*, 548 US 557 126 S.Ct 2749 (29 June 2006) (*Hamdan*)

Akande, D., 'Classification of Armed Conflicts: Relevant Legal concepts' in Elizabeth Wilmshurst (ed.), *International Law and the Classification of Conflicts* (Oxford: Oxford University Press, 2012).

Al-Waheed and Serdar Mohammed v Ministry of Defence [2017] UKSC 2 (*Serdar Mohammed*), available at https://www.supremecourt.uk/cases/docs/uksc-2014-0219-judgment.pdf, accessed 29 October 2018.

Bellinger, J.B. and Padmanabhan, V.M., 'Detention Operations in Contemporary Conflicts: Four Challenges for the Geneva Conventions and Other Existing Law', *American Journal of International Law* 105:2 (2011).

Bethlehem, D., 'The Relationship between International Humanitarian Law and Human Rights Law in Situations of Armed Conflict', *Cambridge Journal of International and Comparative Law* 2:2 (2013).

Çali, B. and Cunningham, S., 'Part 1: A few steps forward, a few steps sideways and a few steps backwards: The CAT's revised and updated GC on Non-Refoulement', *EJIL: Talk!* 20 March 2018, available at https://www.ejiltalk.org/part-1-a-few-steps-forward-a-few-steps-sideways-and-a-few-steps-backwards-the-cats-revised-and-updated-gc-on-non-refoulement/, accessed 29 October 2018.

Chesney, R., Goldsmith, J. and Wittes, B., 'About lawfare: A brief history of the term and the site', *Lawfare Blog*, 1 September 2010, available at https://www.lawfareblog.com/about-lawfare-brief-history-term-and-site, accessed 29 October 2018.

Clapham, A., 'Human Rights Obligations of Non-State Actors', *International Review of the Red Cross* 88:86 (2006).

Coalition Provisional Authority, *Order no 17 (Revised) Status of the Coalition Provisional Authority, MNF-Iraq, Certain Missions and Personnel in Iraq* (Washington DC: Coalition Provisional Authority, 2004).

Council of Europe, *European Convention on Human Rights* (Strasbourg: Council of Europe, 1950) Article 15, available at https://www.echr.coe.int/Documents/Convention_ENG.pdf, accessed 29 October 2018.

Crawford, E., 'Unequal before the law: The case for the elimination of the distinction between international and non-international armed conflicts', *Leiden Journal of International Law* 441 (2007).

Cullen, A., 'Key Developments affecting the Scope of Internal Armed Conflict in International Humanitarian Law', *Military Law Review* 183 (2005).

Danish Ministry of Defence, *The Copenhagen Process on the Handling of Detainees in Military Operations* (Copenhagen: Danish Ministry of Defence, 2012).

Denmark Supreme Court, *Ghousouallah Tarin v Ministry of Defence*, Case no. B1-627-07.

Development, Concepts and Doctrine Centre, *UK Joint Defence Publication 1-10 on Captured Personnel* (London: Ministry of Defence, October 2015), available at https://assets.publishing.service.gov.uk/government/uploads/system/uploads/attachment_data/file/455589/20150820-JDP_1_10_Ed_3_Ch_1_Secured.pdf, accessed 29 October 2018.

Doswald-Beck, L. and Henckaerts, J-M. (eds), *International Humanitarian Law Volume I: Rules* (Cambridge: Cambridge University Press, 2005).

European Court of Human Rights, *Hess v United Kingdom* (1975), available at file:///C:/Users/User/Downloads/001-73854.pdf, accessed 29 October 2018.

European Court of Human Rights, *Loizidou v Turkey, App. No. 15318/89* (Strasbourg: ECtHR, December 1996), available at http://www.asylumlawdatabase.eu/en/content/ecthr-loizidou-v-turkey-application-no-1531889-18-december-1996, accessed 29 October 2018.

European Court of Human Rights, *Sakik and Others v Turkey, App. No. 87/1996/706/898-903* (Strasbourg: ECtHR, November 1997), available at file:///C:/Users/User/Downloads/001-58117.pdf, accessed 29 October 2018.

European Court of Human Rights, *Cyprus v Turkey, App. No. 25781/94* (Strasbourg: ECtHR, May 2001), available at http://www.asylumlawdatabase.eu/en/content/ecthr-cyprus-v-turkey-application-no-2578194-10-may-2001, accessed 28 October 2018.

European Court of Human Rights, *Banković and Others v Belgium and Others, App. No. 52207/99* (Strasbourg: ECtHR, December 2001), available at https://hudoc.echr.coe.int/eng#{%22itemid%22:[%22001-22099%22]}, accessed 29 October 2018.

European Court of Human Rights, *Öcalan v Turkey, App. No. 46221/99* (Strasbourg: ECtHR, May 2005), available at http://www.freedom-for-ocalan.com/english/hintergrund/fall/CASE%20OF%20OCALAN%20v%20TURKEY.pdf, accessed 29 October 2018.

European Court of Human Rights, *Al-Jedda, Behrami and Saramati v France and Others* (*Behrami*) *App. No 71412/01 and 78166/01* (Strasbourg: ECtHR, May 2007), available at https://www.tjsl.edu/slomansonb/3.1_UN_attrib.pdf, accessed 29 October 2018.

European Court of Human Rights, *Medvedyev and Others v France, App. No. 3394/03* (Strasbourg: ECtHR, March 2010), available at file:///C:/Users/User/Downloads/002-1015%20(1).pdf, accessed 28 October 2018.

European Court of Human Rights, *Al-Sadoon and Mufdhi v The United Kingdom App. No. 61498/08* (Strasbourg: ECtHR, March 2010), available at http://www.asylumlawdatabase.eu/en/content/ecthr-%E2%80%93-al-saadoon-and-mufdhi-v-united-kingdom-application-no-6149808-30-june-2009-%E2%80%93, accessed 29 October 2018.

European Court of Human Rights, *Al-Skeini and Others v The United Kingdom, App. No. 55721/07* (Strasbourg: ECtHR, July 2011), available at http://www.asylumlawdatabase.eu/en/content/ecthr-%E2%80%93-al-skeini-and-others-v-united-kingdom-application-no-5572107-7-july-2011, accessed 29 October 2018.

European Court of Human Rights, *Al-Jedda v The United Kingdom App. No. 27021/08* (Strasbourg: ECtHR, 7 July 2011), available at http://www.bailii.org/eu/cases/ECHR/2011/1092.html, accessed 29 October 2018.

European Court of Human Rights, *Hassan v The United Kingdom, App. No. 29750/09* (Strasbourg: ECtHR, September 2014), available at file:///C:/Users/User/Downloads/002-10082.pdf, accessed 29 October 2018.

Geneva Convention relative to the protection of civilian persons in time of war of 12 August 1949, available at http://www.un.org/en/genocideprevention/documents/atrocity-crimes/Doc.33_GC-IV-EN.pdf, accessed 28 October 2018.

Goodman, R., 'The Detention of Civilians in Armed Conflict', *American Journal of International Law* 103:1 (2009).

Greenwood, C., 'Rights at the Frontier: Protecting the Individual in Time of War' in B. Rider (ed.) *Law at the Centre* (Amsterdam: Kluwer Law International 1999).

Hartmann, J., 'The Copenhagen Process: Principles and Guidelines', *EJIL: Talk!* 3 November 2012, available at https://www.ejiltalk.org/the-copenhagen-process-principles-and-guidelines/, accessed 29 October 2018.

Horowitz, J., 'Interlocutory note to the Copenhagen Process, Principles and Guidelines on the handling of detainees in military operations', *ASIL International Legal Materials* (2012), available at https://papers.ssrn.com/sol3/papers.cfm?abstract_id=2228372, accessed 29 October 2018.

International Committee of the Red Cross, *Expert meeting on procedural safeguards for security detention in non-international armed conflict* (London: ICRC, 2009), available at https://www.icrc.org/eng/assets/files/other/irrc-876-expert-meeting.pdf, accessed 29 October 2018.

International Court of Justice, *Legal Consequences of the Construction of a Wall in the Occupied Palestinian Territories* (The Hague: ICJ, 2004), 136, available at http://www.unrod.org/docs/ICJ-Advisory2004.pdf, accessed 29 October 2018.

International Court of Justice, *DRC v Uganda*, available at https://www.asil.org/insights/volume/10/issue/1/case-concerning-armed-activities-territory-congo-icj-finds-uganda-acted, accessed 29 October 2018.

International Criminal Court, *The Prosecutor v Jean-Pierre Bemba*, ICC-01/05-01/08-424, Confirmation of Charges Decision (Pre-Trial Chamber), 15 June 2009, available at https://www.icc-cpi.int/car/bemba, accessed 29 October 2018.

International Criminal Database, Maya Evans case, available at http://www.internationalcrimesdatabase.org/Case/1194, accessed 29 October 2018.

International Criminal Tribunal for the Former Yugoslavia, *Prosecutor v Dusko Tadić (jurisdiction of the Tribunal)* ICTY case no. IT-94-1-AR72 (2 Oct 1995), available at http://www.icty.org/case/tadic/4, accessed 29 October 2018.

International Criminal Tribunal for the Former Yugoslavia, *Prosecutor v Ramush Haradinaj* ICTY case no. IT-04-84-T (3 April 2008), available at http://www.icty.org/x/cases/haradinaj/tjug/en/080403.pdf, accessed 29 October 2018.

International Law Commission, *Report on the Draft Articles on the Responsibilities of International Organisations Report 58th Session* (New York: United Nations, 2006), available at http://legal.un.org/docs/?path=../ilc/reports/2006/english/chp7.pdf&lang=EFSRAC, accessed 29 October 2018.

International Law Commission, *Fragmentation of International Law: Difficulties arising from the Diversification and Expansion of International Law: Report of the Study Group of the International Law Commission* (New York: United Nations, 2006), available at http://legal.un.org/ilc/documentation/english/a_cn4_l682.pdf, accessed 29 October 2018.

Milanović, M. and Papić, T., 'As bad as it gets: The European Court of Human Right's Behrami and Saramati Decision and General International Law', *International and Comparative Law Quarterly* 58 (2009).

Pejić, J., 'Procedural Principles and Safeguards for Internment/Administrative Detention in Armed Conflict and Other Situations of Violence', *International Review of the Red Cross* 87:858 (2005).

Protocol Additional to the Geneva Conventions of 12 August 1949, and relating to the Protection of Victims of Non-International Armed Conflicts (Protocol II) in 8 June 1977 Art 6, available at https://ihl-databases.icrc.org/ihl/ INTRO/475?OpenDocument, accessed 29 October 2018.

Roberts, A. and Guelff, R. (eds), *Documents on the Laws of War* (Oxford: Oxford University Press, 2001).

Rodley, N. and Pollard, M., *The Treatment of Prisoners under International Law* (Oxford: Oxford University Press, 2009).

Rowe, G., 'Is there a right to detain civilians by foreign armed forces during a non-international armed conflict?', *International and Comparative Law Quarterly* 61:3 (2012).

Sassóli, M., 'The International Legal Framework for Stability Operations: When May International Forces Attack or Detain Someone in Afghanistan?', *International Law Study Series* 85 (2009).

Sivakumaran, S., 'Binding Armed Opposition Groups', *International and Comparative Law Quarterly* 55:2 (2006).

Sivakumaran, S., *The Law of Non-International Armed Conflict* (Oxford: Oxford University Press, 2012).

Sivakumaran, S., 'IHL Challenges series: Typology of Conflicts Part IV: Who decides?', *ICRC Blog* 28 March 2013, available at http://intercrossblog.icrc. org/blog/ihl-challenges-series-typology-of-conflicts-part-iv-who-decides, accessed 29 October 2018.

Solf, W., 'Problems with the application of norms governing interstate armed conflict to non-international armed conflict', *Georgia Journal of International and Comparative Law* 13 (1983).

Solis, G., *The Law of Armed Conflict* (Cambridge: Cambridge University Press, 2010).

The Human Rights Committee, General Comment No. 29 (New York: United Nations, 2001), available at http://hrlibrary.umn.edu/gencomm/hrc29.html, accessed 28 October 2018.

United Nations, *United Nations Security Council Resolution 1386 on the situation in Afghanistan on 20 December 2001* (New York: United Nations, 2001).

United Nations, 'Military Technical Agreement between ISAF and the Interim Administration of Afghanistan' (New York: United Nations, January 2002) Annex A, para. 3, available at https://reliefweb.int/report/afghanistan/ afghanistan-militarytechnical-agreement-between-international-security, accessed 29 October 2018.

UK Ministry of Defence, *The Joint Service Manual of the Law of Armed Conflict JSP 383* (London: Ministry of Defence, 2004), available at https://assets.publishing. service.gov.uk/government/uploads/system/uploads/attachment_data/ file/27874/JSP3832004Edition.pdf, accessed 29 October 2018.

UK Supreme Court, *Smith (No 2) et al. v Ministry of Defence [2013] UKSC 41* (London: House of Lords, 2013), available at https://www.supremecourt.uk/decided-cases/docs/UKSC_2012_0249_Judgment.pdf, accessed 29 October 2018.

UK Supreme Court, *Al-Waheed and Serdar Mohammed v Ministry of Defence* [2017] UKSC 2 (*Serdar Mohammed*), available at https://www.supremecourt.uk/ cases/docs/uksc-2014-0219-judgment.pdf, accessed 29 October 2018.

United Nations Committee against Torture, *General Comment No. 1 (2017) on the Implementation of Article 3 of the Convention in the Context of Article 22* (New York: United Nations, February 2017), available at https://www.ohchr.org/ Documents/HRBodies/CAT/GCArticle3/CAT-C-GC-1.pdf, accessed 29 October 2018.

United Nations, *Criminal Accountability of United Nations Officials and Experts on Mission* (New York: United Nations, September 2007), available at http:// legal.un.org/docs/?symbol=A/62/329, accessed 28 October 2018.

United Nations, *The Secretary-General's High-level Panel report on Threats, Challenges and Change: A more secure world, our shared responsibility* (New York: United Nations, 2004), available at https://www.un.org/ruleoflaw/files/gaA.59.565_ En.pdf, accessed 29 October 2018.

United Nations, *Final Record of the Diplomatic Conference of Geneva of 1949 Vol III* (New York: United Nations, 1949), available at https://www.loc.gov/rr/frd/ Military_Law/pdf/Dipl-Conf-1949-Final_Vol-3.pdf, accessed 29 October 2018.

United Nations Human Rights Committee, *General Comment 31: The Nature of the General Legal Obligation Imposed on States Parties to the Covenant* (Geneva: United Nations, May 2004), available at https://www.ohchr.org/EN/ Issues/Education/Training/Compilation/Pages/c)GeneralComment No31TheNatureoftheGeneralLegalObligationImposedonStatesPartiestothe Covenant(2004).aspx, accessed 29 October 2018.

United Nations, *International Covenant on Civil and Political Rights* (New York: United Nations, 1966), available at http://www.un-documents.net/iccpr.htm, accessed 29 October 2018.

US Department of State, *United States Second and Third Periodic Report of the United States of America to the UN Committee on Human Rights Concerning the International Covenant on Civil and Political Rights* (Washington DC: State Department, October 2005), available at https://www.state.gov/j/drl/ rls/55504.htm, accessed 29 October 2018.

Winkler, T., 'The Copenhagen Process on Detainees: A Necessity', *Nordic Journal of International Law* 78 (2010).

PART FOUR

THE UK DOMESTIC CONTEXT OF 'WAR AMONGST THE PEOPLE'

9

UK DOMESTIC SUPPORT FOR WARS

Polls and Public Opinion

Ian Wilson

In this age, in this country, public sentiment is everything. With it, nothing can fail; against it, nothing can succeed. Whoever moulds public sentiment goes deeper than he who enacts statutes, or pronounces judicial decisions.[1]

Abraham Lincoln

My great fear is that we as a nation will become so risk-averse, so cynical and so introverted that we will find ourselves in inglorious and impotent isolation by default.[2]

Bill Rammell (Minister for the Armed Forces, 2009–10)

Introduction

When the phrase 'war amongst the people' is referred to, generally it is not the intervening state's 'people' that are being talked about. This is a mistake. Domestic opinion matters, yet is largely misunderstood and often misrepresented. This chapter is about understanding contemporary UK public opinion in times of war. Abraham Lincoln and Bill Rammell, despite being politicians from different eras, both articulated the importance of 'winning' the public. In democracies, surveys and polls are a well-established way of assessing public opinion on a number of matters; Franklin D. Roosevelt used polling in the two years before Pearl Harbor to

---

1 See Abraham Lincoln, *Collected Works of Abraham Lincoln Volume Two* (Rockville: Wildside Press, 2008).

2 'Risk averse Britain may lose stomach for war, warns minister', *The Times* 12 January 2010, available at https://www.thetimes.co.uk/article/risk-averse-britain-may-lose-stomach-for-war-warns-minister-dbw7lnbbn0x, accessed 25 October 2018.

gauge the public mood for war.[3] In short, public opinion matters and polls remain (despite their noted imperfections) the best mechanism to assess and understand it. Crucially, from a policy perspective, understanding the way that the public mood is determined might also help identify how it could be influenced or shaped. Public opinion is viewed as a potential weakness in government policymaking, with phrases like the 'Shadow of Iraq' or the 'Vietnam Syndrome' used to explain past events as constraints on public support for new wars.[4] Indeed, the Ministry of Defence's concern was such that, in November 2012, a report was written effectively asking the question 'how can we better persuade the public to support a UK war?'[5] Despite this clear interest, little empirical study has been done on such matters from a UK perspective.

This chapter offers a better understanding of how polls can help 'read' the British public and where they fit into the paradigm of 'war amongst the people'. The first part will outline some of the existing theories and evidence on public opinion and war, grouping them into three broad theoretical groups or schools of thought. For each group, a hypothesis is proposed and tested against contemporary case studies. The chapter is split into two substantive sections: first outlining and then analysing the findings. The outline of findings begins by explaining how the case studies are affected by differing levels of scale, intensity and duration and then unpacks the categories of public support, perception of success and prime ministerial approval. These categories then provide the framework for the analysis. The second section tests the polling data against the three theoretical models identified in section one: the *Ends and Means Calculus*, the *Principle Policy Aim* and the *Leadership Effect* theory. The chapter's core argument is that the British public seems to make an assessment based on the likelihood of success balanced against the perceived legitimacy of the aim. This matters when discussing the concept of 'war amongst the people', because this constituency is now, more than ever, a key audience.

---

3 See Thomas Scotto, Jason Reifler, Harold Clarke, Julio Lopez, David Sanders, Marianne Stewart and Paul Whiteley, *Briefing Paper: Attitudes towards British Involvement in Afghanistan* (Essex: Institute for Democracy and Conflict Resolution 2011).

4 Peter Kellner, 'Syria and the long shadow of Iraq', *Daily Telegraph* 28 August 2013, available at https://www.telegraph.co.uk/news/worldnews/middleeast/syria/10270344/Peter-Kellner-Syria-and-the-long-shadow-of-Iraq.html, accessed 25 October 2018.

5 Development, Concepts and Doctrine Centre, *Risk: The Implications of Current Attitudes to Risk for the Joint Operational Concept* (Shrivenham: Ministry of Defence, 2012); Ben Quinn, 'MoD study sets out how to sell wars to the public', *The Guardian* 26 September 2013, available at https://www.theguardian.com/uk-news/2013/sep/26/mod-study-sell-wars-public, accessed 25 October 2018.

Existing Theories and Evidence

Scholarship is divided on what constitute the most important factors affecting public support for war. Despite this, a number of mainstream theories emerge from the existing literature. There are, broadly speaking, three groups with some overlap between them, as some explanations do not fit neatly into one school.[6] In recent years, there has been an increase in the number of cross-theory and multidimensional studies and, while different theorists use slightly different terminology, the underpinning ideas are generally consistent.[7]

The Ends and Means Calculus

The first of these is built upon an 'ends and means calculus'. Proponents of this theory argue that the public performs cost-benefit calculations when deciding whether to support military action.[8] The theory was first popularised in 1996 by Eric Larson in *Casualties and Consensus*, where he examines polling data relating to US twentieth-century conflicts – from the Second World War through to Somalia – and argues that the public will support wars where the costs are perceived to be worth the benefits. Larson bases his calculation of ends and means on three parameters. First, there are the expected costs where public opinion tends to develop in inverse proportion to the actual costs.[9] In this model, the cost to the public tends to

---

6   While the three main theories or schools outlined are useful, they are not exhaustive. It would be over-simplistic to try and explain public support for war by focusing on only a single variable, leaving out theories that do not fit neatly into the mainstream categories. With multidimensional approaches becoming increasingly popular, there is a growing acceptance that a 'golden' or Unified Theory of Public Opinion is unlikely to emerge.

7   Ebru Canan-Sokullu, 'Domestic Support for Wars: A Cross-Case and Cross-Country Analysis', *Armed Forces and Society* 38:38 (2011), 117–41; Ben Clements, 'Public Opinion and Military Intervention: Afghanistan, Iraq and Libya', *The Political Quarterly* 84:1 (2013), 119–31; and Jason Reifler, Harold Clarke, Thomas Scotto, David Sanders, Marianne Stewart and Paul Whiteley, 'Prudence, Principle and Minimal Heuristics: British Public Opinion toward the Use of Military Force in Afghanistan and Libya', *British Public Opinion* 16:1 (2014), 28–55.

8   See Eric Larson, *Casualties and Consensus: The Historical Role of Casualties in Domestic Support for US Military Operations* (Santa Monica: RAND, 1996); Richard Hermann, Phillip Tetlock and Penny Visser, 'Mass public decisions to go to war: A cognitive-interactionist approach', *American Political Science Review* 93:3 (1999), 553–73; Christopher Gelpi, Peter Feaver and Jason Reifler, 'Success Matters: Casualty Sensitivity and the War in Iraq', *International Security* 30:3 (2006), 7–46; and Christopher Gelpi, Peter Feaver and Jason Reifler, *Paying the Human Costs of War: American Public Opinion and Casualties in Military Conflicts* (New Jersey: Princeton University Press, 2009).

9   Steven Kull, 'Review of Eric Larsen's *Casualties and Consensus*', *Public Opinion Quarterly* 61 (1997), 687–722 and Larson, *Casualties and Consensus*.

be measured in combat fatalities and casualty sensitivity – the public's price sensitivity to the human cost of war. Second, there are the perceived benefits of military intervention, which includes diffuse areas such as principles or values, economic interest and strategic national interests. Moreover, the theory suggests that the greater the public interest in the outcome (or ends) the more willing they will be to support military actions (and the attendant costs). The third parameter is the 'prospect for success'. Success is necessarily time critical and Larson notes that 'operations failing to achieve objectives quickly tend to lose support'.[10] However, if success occurs more quickly than expected, public support is likely to rise commensurately.

Critics of the 'ends and means' school of thought argue that the model ascribes too much agency and rationality to the public. They suggest that most citizens are not capable of making such cost-benefit calculations, relying instead on heuristics or readily available cues from political elites, an area covered in the 'Leadership Effect' section below.[11] Another criticism is that the literature, while ostensibly about both aspects, focuses more heavily on costs. Detractors argue that casualty sensitivity is often exaggerated, which, in turn, leads to false lessons being learned. It is worth noting that there is a distinction between casualty sensitivity and casualty phobia and, when war aims are vague, multifaceted or constantly changing, it can be difficult to establish an effective 'benefit' metric.[12] If correct, the policy implications are that politicians ought to select wars that the UK public view as both worthwhile and winnable, while weighing up the expected costs. At its simplest, this is an ends, ways and means calculus. Therefore, Hypothesis one (H1) proposes that domestic support for wars is sensitive to both the 'costs' and the public's perception of 'success'.[13]

The Principle Policy Aim

The second mainstream theory to be considered is the 'principle policy aim', which emphasises the importance of morality, including concepts

---

10 Larson, *Casualties and Consensus*, 180.

11 Matthew Baum and Phillip Potter, 'The Relationships Between Mass Media, Public Opinion and Foreign Policy: Towards a Theoretical Synthesis', *Annual Review of Political Science* 11 (2008), 39–65; Adam Berinsky and James Druckman, 'Public Opinion research and support for the Iraq War', *Public Opinion Quarterly* 71:1 (2007), 126–41.

12 Gary Jacobsen, 'A Tale of Two Wars: Public Opinion on the US Military Interventions in Afghanistan and Iraq, *Presidential Studies Quarterly* 40:4 (2010), 585–610; John Mueller, 'Trends in Popular Support for the Wars in Korea and Vietnam', *American Political Science Review* 65:2 (1971), 358–75 and Christopher Gelpi, Peter Feaver and Jason Reifler, *Paying the Human Costs of War*.

13 I have deliberately omitted testing 'perceived benefits' because, in the case studies used in this chapter, they are extremely hard to quantify, let alone measure.

such as 'just war' theory and 'legitimacy', suggesting that the public's attitude towards war is shaped by the expressed aim, which is ascribed some kind of moral weight. It can also be expressed in terms of perceived interest and is not as binary as 'right' or 'wrong'. Bruce Jentleson lists three main types of 'principle policy objectives', categorised as 'foreign policy restraint', 'internal political change' and 'humanitarian intervention'. The foreign policy restraint objective is where military means are used to change the behaviours of another state that has used aggression.[14] His second objective – 'internal political change'– is an effort to alter the political system of another state. This can be manifested not only through 'regime change' but also by other methods, including insurgency or terrorism. Lastly, 'humanitarian intervention' is where external powers intervene with the purpose of 'relieving grave human suffering' or to 'rescue people at risk from political causes'.[15] This category, which is contested, provides a moral basis to override the sovereignty of transgressing regimes and lays the theoretical groundwork for the Responsibility to Protect (R2P) doctrine. If this theory is correct, leaders should choose wars that the UK public see as both just and legitimate. Consequently, securing UN approval and a wider international consensus become much more important. Hypothesis two (H2) therefore notes that domestic support for wars depends on the public viewing war as moral and legitimate and/ or in the national interest.

The Leadership Effect Theory

The 'leadership effect' theory argues that popular and likeable leaders can influence public opinion through affective heuristics. Examples within the literature include leadership popularity,[16] elite cues[17] and the rally-round-the-flag effect,[18] with debate focusing on how effectively these cues impact on public opinion. Berinsky offers a pessimistic account of cue-giving,

---

14   Bruce Jentleson, 'The Pretty Prudent Public: Post-Vietnam American Opinion on the Use of Military Force', *International Studies Quarterly* 36:1 (1992), 49–74.

15   Jentleson, 'The Pretty Prudent Public', 53.

16   Julie Karbo, 'Prime Minister Leadership Styles in Foreign Policy Decision making: A Framework for Research', *Political Psychology* 18:3 (1997), 553–81.

17   Paul Baines and Robert Worcester, 'When the British "Tommy" went to war, public opinion followed', *Journal of Public Affairs* 5 (2005), 4–19; John Zaller, *The Nature and Origins of Mass Opinion*, (Cambridge: Cambridge University Press, 1992).

18   William Baker and John O'Neal, 'Patriotism or opinion leadership? The nature and origins of the 'rally round the flag' effect', *Journal of Conflict Resolution* 44:5 (2001), 661–87; Matthew Baum, 'The constituent foundations of the rally-round-the-flag phenomenon', *International Studies Quarterly* 46:2 (2002), 263–98.

suggesting that it can be highly partisan, dividing on ethnic or gender lines.[19] Others refer to the 'halo' effect, where perceptions of traits such as attractiveness and charisma lead to the conclusion that politicians are more trustworthy than they actually are.[20] The wider claim of the theory, however, is that heuristics is a more powerful determinant of public opinion, with emotion, not rationality, generally dominant.[21] If correct, this would mean that leaders and their approval ratings could help explain variations in public support for wars.

A common criticism of this theory is that leaders may be following public opinion rather than shaping it. Of course, this depends on the type of leadership and the role foreign policy is perceived to play in elections. Baum and Porter argue for a flatter, less hierarchical model, where there is a 'marketplace' of opinion where the public, the elite and the media all interact and influence each other to varying degrees at different times.[22] Others argue that the 'leadership effect' is more limited, with effectiveness depending on *which* elites are 'framing' the policy and *when* they do so.[23] If this theory is correct, politicians with high approval ratings have a better chance of persuading the public to support a war. Moreover, it suggests that leaders should embark on wars when their approval ratings are highest, so timing in the electoral cycle would be important. Hypothesis three (H3) posits that domestic support for wars can be influenced by a leader with a high public approval rating.

Scale, Intensity and Duration

The three conflicts covered in this chapter – Afghanistan, Iraq and Libya – need to be considered in terms of scale, intensity and duration. The Syrian crisis is distinctive in that, from a British perspective, it was a 'phoney

---

19 Berinsky and Druckman. 'Public Opinion research and support for the Iraq War', 126–41.

20 Jeremy Dean, 'The Halo Effect: When Your Own Mind is a Mystery'. *PsyBlog* 31 October 2007, available at http://www.spring.org.uk/2007/10/halo-effect-when-your-own-mind-is.php, accessed 25 October 2018; Kevin Murphy, Robert Jako and Rebecca Anhalt, 'Nature and consequences of halo error: A critical analysis', *Journal of Applied Psychology* 78:2 (1993), 218–25.

21 James Kuklinski and Paul Quirk, 'Reconsidering the rational public: Cognition, heuristics and mass opinion' in Arthur Lupia, Matthew McCubbins and Samuel Popkin (eds), *Elements of Reason* (Cambridge: Cambridge University Press, 2000).

22 Baum and Potter, 'The Relationships between Mass Media, Public Opinion and Foreign Policy', 39–65.

23 Adam Berinsky and Donald Kinder, 'Making Sense of Issues through Media Frames: Understanding the Kosovo Crisis' *The Journal of Politics* 68:3 (2006), 640–56.

intervention', at least initially.[24] Both Iraq and Afghanistan were broadly comparable in terms of scale, with troop numbers ranging from 8,000–10,000 at the peak of combat operations in each intervention.[25] During the Blair premiership, fewer British troops were deployed to Afghanistan than Iraq. However, when military operations in Iraq ended in April 2009, much of the troop dividend was reinvested in Afghanistan. Libya was much smaller in scale than either Iraq or Afghanistan; at its peak, approximately 1,300 personnel were involved.[26] Scale is important because it informs the expectation of casualty rates.[27]

In terms of intensity – defined as how hard the fighting is, with casualty numbers the best measure[28] – Afghanistan was the most deadly and averaged three casualties per month to Iraq's 2.7.[29] In Iraq, there was a spike in casualties during the initial invasion of 2003; injuries then fell to an average of less than 10 per month. This meant that, until August 2007, the monthly casualty rate in Iraq was higher than Afghanistan, which remained at less than 5 per month[30] (see Figure 9.1). However, from late 2007, British casualties in Iraq fell sharply following the transfer of control in Basra to Iraqi forces. Ostensibly, both casualty levels and troop numbers in Afghanistan peaked through 2009–10.[31] However, on closer examination, the monthly average in Afghanistan actually rose to almost 5 (4.7) per month during the UK 'surge' into Helmand Province in 2005–2014.[32] Meanwhile, Iraq's figure dropped to less than two (1.9) when the initial invasion is discounted.[33] Overall, the UK suffered more fatalities in Afghanistan (453) than Iraq (179) (see Figure 9.2). There were no combat casualties in the Libya campaign. This is particularly important when assessing the ends and means calculus (H1).

---

24 This refers to the proposed direct intervention in Syria (bombing or non-SF land forces), not involvement in the wider region.

25 'UK Troops in Afghanistan', *BBC News* (2014) available at https://newsimg.bbc.co.uk/media/images/46830000/gif/_46830596_afg_troops_226_3.gif, accessed 25 October 2018.

26 'NATO operations in Libya: data journalism breaks down which country does what', *The Guardian* 22 May 2011 available at http://www.theguardian.com/news/datablog/2011/may/22/nato-libya-data-journalism-operations-country#_, accessed 25 October 2018.

27 Gelpi, Feaver and Reifler, *Paying the Human Costs of War*.

28 For clarity casualties, fatalities, and deaths are considered to be the same thing. I do not use battle injuries or wounding in the statistics.

29 There were 179 casualties over 79 months, which comes to a 2.7 average.

30 The spike in September 2006 was due to an air accident in Afghanistan involving a Nimrod MR2 aeroplane, where 14 servicemen were killed.

31 'UK Troops in Iraq', *BBC News* (2009), available at https://newsimg.bbc.co.uk/media/images/42597000/gif/_42597107_uk_troop_numbers203.gif, accessed 25 October 2018.

32 There were 449 casualties over 96 months, which comes to a 4.7 average.

33 There were 152 casualties over 78 months, which comes to a 1.9 average – as 27 were killed in the initial ground invasion.

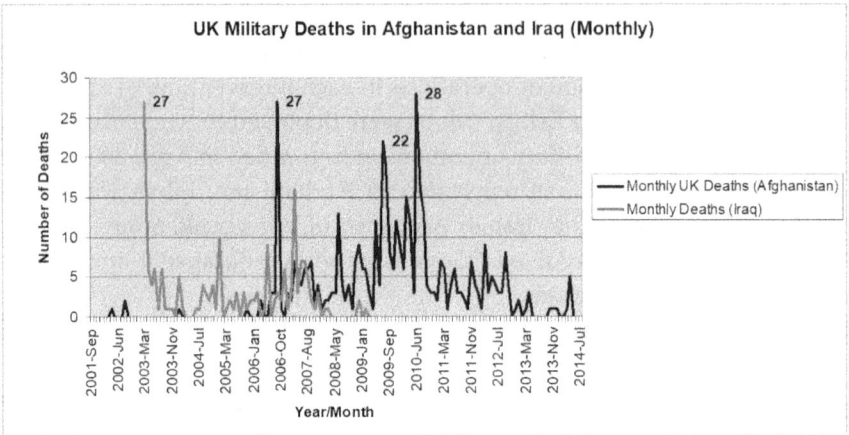

Figure 9.1   UK Military Fatalities in Iraq and Afghanistan, 2001–14 (Monthly)

Sources: BBC News and Casualty Monitor[34]

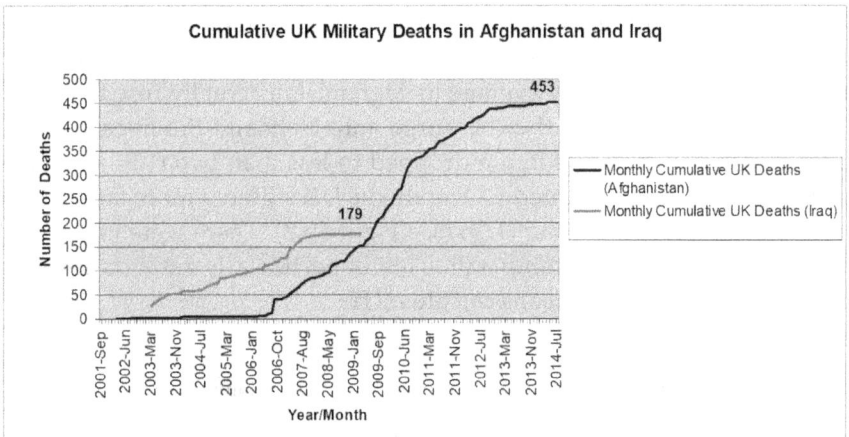

Figure 9.2   UK Military Fatalities in Iraq and Afghanistan, 2001–14 (Cumulative)

Sources: BBC News and Casualty Monitor[35]

---

34 'UK Military Deaths in Iraq', *BBC News* 7 July 2016, available at http://www.bbc.co.uk/news/uk-10637526?print=true, accessed 25 October 2018; 'UK Military Deaths in Afghanistan', *BBC News* 12 October 2015, available at http://www.bbc.co.uk/news/uk-10629358?print=true, accessed 25 October 2018; 'British Casualties: Afghanistan' (2016) *Casualty Monitor*, available at http://www.casualty-monitor.org/p/british-casualties-in-afghanistan.html, accessed 25 October 2018; 'British Casualties: Iraq' *Casualty Monitor* 31 July 2009, available at http://www.casualty-monitor.org/p/iraq.html, accessed 25 October 2018.

35 *BBC News*, 'UK Military Deaths in Iraq'; *BBC News*, 'UK Military Deaths in Afghanistan'; *Casualty Monitor*, 'British Casualties: Afghanistan'; *Casualty Monitor*, 'British Casualties: Iraq' (2009).

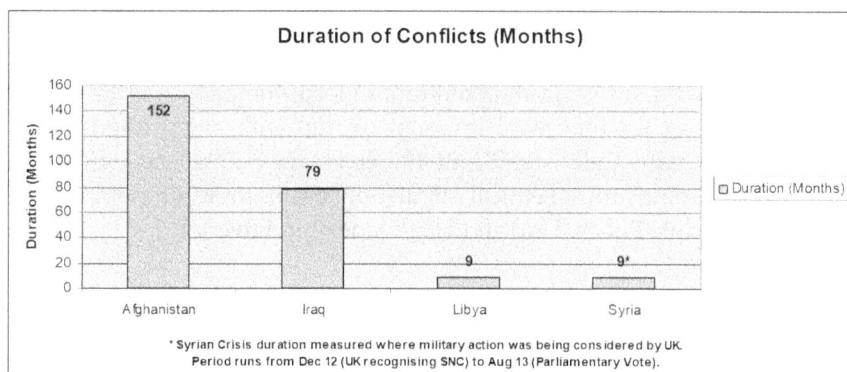

Figure 9.3   **Duration of UK Involvement in Wars/Crises**
Source: Data collated by the author

Afghanistan was by far the longest of the case studies, with the UK having a combat commitment from October 2001 to October 2014. In contrast, the campaign in Iraq ran from March 2003 until June 2009, a period of 79 months. Libya was even more limited, lasting only 9 months (see Figure 9.3). British involvement in Iraq lasted longer than the First World War, while Afghanistan has been longer than the First and Second World Wars combined. Again, duration is a factor in means and ends calculus (H1) as time (and money) could influence public views on whether or not to support a conflict.

Public Support

In the wake of 9/11, the British public strongly backed the intervention in Afghanistan. Polls taken at the end of 2001 indicated that 65 per cent of Britons supported joining a US-led coalition.[36] Contrary to popular belief, a slight majority (53 per cent) of Britons also supported the invasion of Iraq. However, this was caveated with the need for it to be 'approved by the United Nations'; without this, support dropped to only 13 per cent.[37] With Libya, public opinion at the outset was finely balanced, with one poll suggesting that 37 per cent supported air strikes, 36 per cent were opposed and 28 per cent were undecided.[38] Initial support is a good barometer of how much the

---

36 Scotto, Reifler, Clarke, Lopez, Sanders, Stewart and Whiteley, *Briefing Paper: Attitudes towards British Involvement in Afghanistan*, 2.

37 'YouGov/ITN on the justification of war in Iraq', *YouGov* 10–12 January 2003, available at http://cdn.yougov.com/today_uk_import/YG-Archives-Ira-itn-WarIraq-030113.pdf, accessed 25 October 2018.

38 Clements, 'Public Opinion and Military Intervention: Afghanistan, Iraq and Libya', 122.

'rally-round-the-flag' effect is in play, especially if there is a subsequent drop in support, and also provides some insight into the public's perception of the rightness or merit of an intervention. Indeed, the margin of difference between support for the initial invasion of Iraq with and without a UN resolution is a statistically significant 40 percentage points (see Figure 9.4). This suggests some form of ethical calculation, useful for testing elements of both H2 (Principle Policy Aim) and H3 (Leadership Effect).

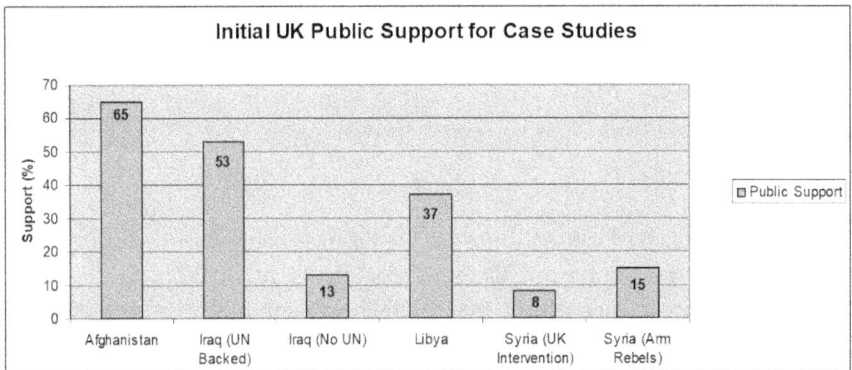

Figure 9.4   Public Support at the Outset of Wars/Crises
Sources: Data collated by the author

Public support for the war in Iraq declined relatively steadily, dropping permanently below 50 per cent in February 2004 and hitting the nadir of 19 per cent in August 2007 (see Figure 9.5). Unfortunately, data from YouGov charting public support beyond 2007 is unavailable. In Afghanistan, 2009 was the public opinion tipping point, with a BBC *Newsnight* poll indicating that 46 per cent supported UK involvement and 47 per cent opposed it,[39] while a survey by Jason Reifler in February 2009 indicated an even greater disparity, with only 22 per cent approving and 51 per cent against.[40] This shift in the public mood was confirmed by an *Independent* poll in July of the same year, where 52 per cent wanted the UK to withdraw immediately, compared to 43 per cent in favour of remaining.[41] It is worth noting that

---

39 Julian Glover, 'Survey of public opinion on Afghan conflict finds support and doubt casts', *The Guardian* 13 July 2009, available at https://www.theguardian.com/uk/2009/jul/13/public-opinion-poll-afghanistan-war, accessed 25 October 2018.

40 Reifler, Clarke, Scotto, Sanders, Stewart and Whiteley, 'Prudence, Principle and Minimal Heuristics', 36.

41 Kim Sengupta and Nigel Morris, 'Voters turn against war in Afghanistan', *The Independent* 28 July 2009, available at http://www.independent.co.uk/news/uk/politics/voters-turn-against-war-in-afghanistan-1763227.html, accessed 25 October 2018.

UK Public Support for War in Iraq

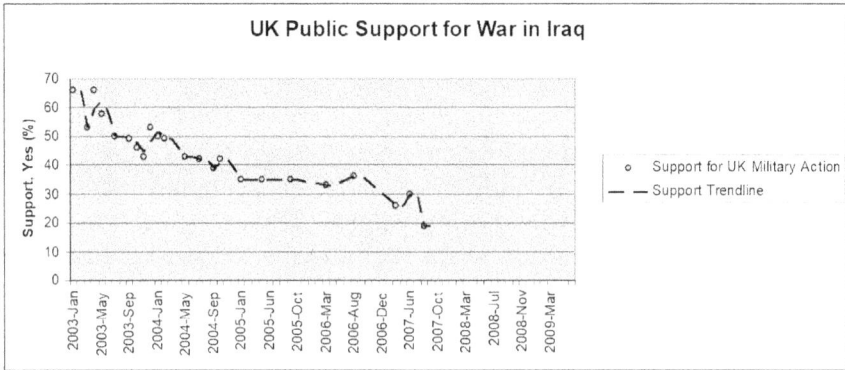

Figure 9.5   Public Support for Military Action in Iraq

Sources: Author from YouGov polls[42]

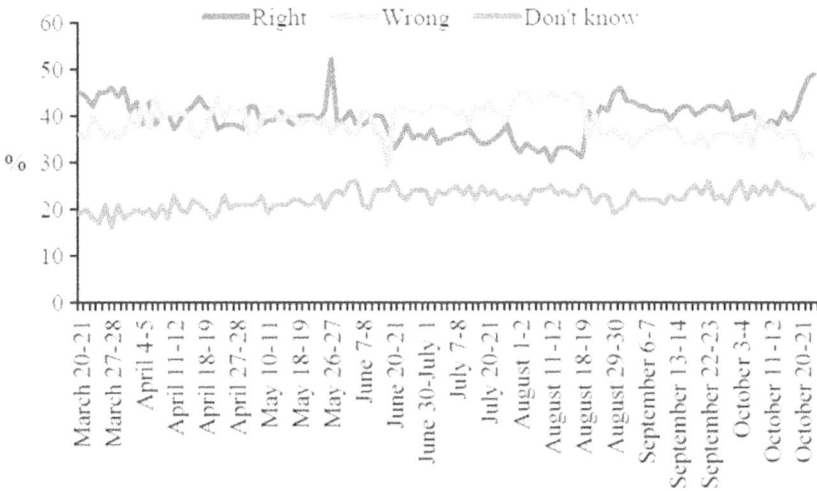

Figure 9.6   Public Support for Military Action in Libya

Question: 'Do you think Britain, France, the US and other countries are right or wrong to take military action in Libya?'

Source: Clements 2013 from YouGov polls[43]

---

42 'Iraq War: Wrong or Right?' *YouGov* (2013), available at http://cdn.yougov.com/cumulus_uploads/document/raghpsamv0/YG-Archives-Pol-Trackers-Iraq-130313.pdf, accessed 25 October 2018.

43 Clements, 'Public Opinion and Military Intervention: Afghanistan, Iraq and Libya', 128.

2009 was the year when the number of British fatalities in Afghanistan surpassed the Iraq totals. However, trends are harder to extrapolate as polls were conducted less frequently than in Iraq.

With Libya, YouGov polling revealed a greater level of fluctuation. The trend lines often run in close proximity through the months of April and May, while in June and July the numbers believing the intervention was wrong exceed those who thought it right. The numbers reversed from August onwards before a final sharp rise in October that coincided with Colonel Gaddafi's death (see Figure 9.6).

Perception of Success

Measuring the success of a campaign – as opposed to how well it is supported – is where the polling data is most scarce. YouGov has produced useful data for a compressed period of time within the Afghanistan campaign but it does not cover eleven of the thirteen years of UK involvement. As a snapshot, it shows that there was a significant rise (from 42 per cent to 53 per cent) in the 'No – not possible' response category when British casualties peaked at 28 in June 2010 (see Figure 9.7). After that, it remained at approximately 50 per cent, save a drop to 44 per cent in September 2011 and a spike of 63 per cent in March 2012.[44] It is possible that the 'No – not possible' response captures (in line with H1 on ends and means) the impact of a particularly bad month. Perhaps the most telling trend is the steady fall of those who considered the more optimistic 'no – but will eventually', where, over a two-year period, there is a smaller drop from 33 per cent to only 16 per cent. When allied with the 'No – not possible' camp, a steady but significant divergence can be discerned: from a 10-point gap (33 per cent and 43 per cent) in April 2010 to a 43-point delta (16 per cent and 59 per cent) in April 2012 (Figure 9.7). In the case of Libya, the data is better, as YouGov tracked public attitudes throughout the build-up to and prosecution of the campaign. Figure 9.8 notes the fluctuation in perceptions of success over time. As a limited air campaign, there are no UK casualties to test for correlation. The trends suggest that the public mood may be linked both to military developments in theatre and how these are portrayed by the media and political leaders, with sharp rises in positive perceptions in August corresponding with tactical successes on the ground, including the fall of Tripoli.[45] The final spike in October (48 per cent) relates to the end of hostilities and the capture and killing of Gaddafi.

---

44 The nine casualties in March 2012 was the highest total in 17 months and may have affected the spike in the 'No – not possible' camp.

45 Clements, 'Public Opinion and Military Intervention: Afghanistan, Iraq and Libya', 128.

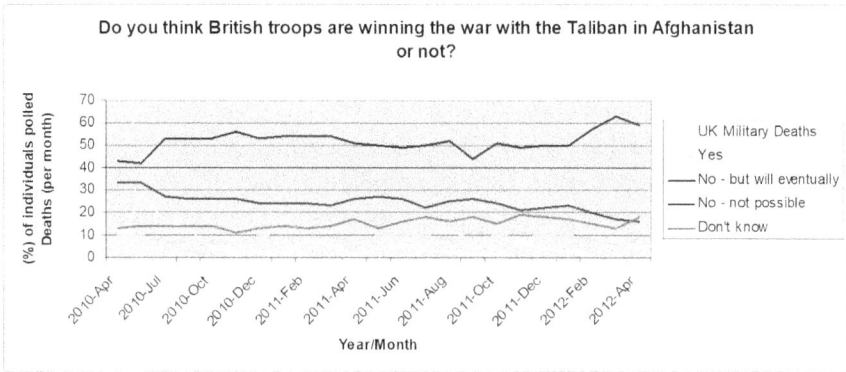

Figure 9.7   Public Perception of Success in Afghanistan

Source: Author from (YouGov 2012b) polls[46]

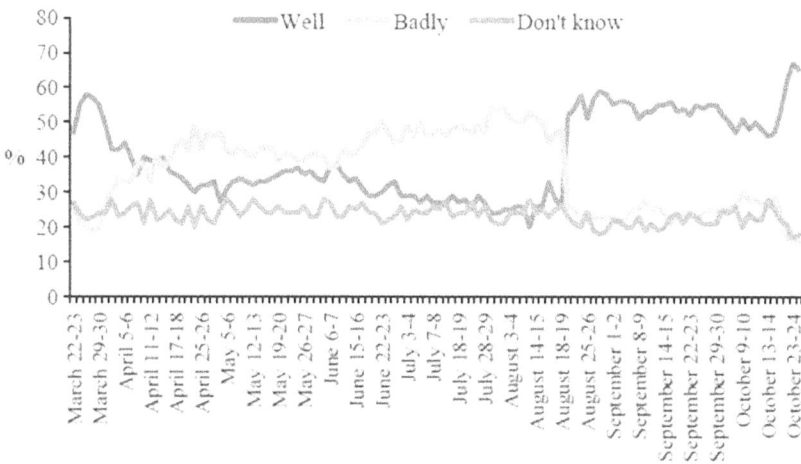

Figure 9.8   Public Perception of Success in Libya

Question: 'Overall, do you think the Coalition's military action in Libya is going well or badly?'

Source: Clements 2013 from YouGov polls[47]

---

46 'Afghanistan Tracker' *YouGov* (2012), available athttp://d25d2506sfb94s.cloudfront.net/cumulus_uploads/document/m36o16yfbo/YG-Archives-Pol-Trackers-Afghanistan-240412.pdf, accessed 25 October 2018.

47 Clements, 'Public Opinion and Military Intervention: Afghanistan, Iraq and Libya', 128.

Prime Ministerial Approval

Finally, the popularity of a Prime Minister may affect or be affected by their foreign policy decisions with regard to war. PM approval ratings may reveal a link to the Leadership theory (H3). Figure 9.9 illustrates that PMs are, on average, more popular than their governments, with a noted 'bounce' in approval ratings post-election or after a leadership change. Anything more than a 4 per cent monthly variation is statistically significant and indicates a shift in public perception, though not necessarily influenced by foreign policy. Moreover, significant gaps between leadership and government approval might indicate a 'halo' or 'negative halo' effect. The Ipsos-Mori data is extensive and provides an accurate reading of monthly approval

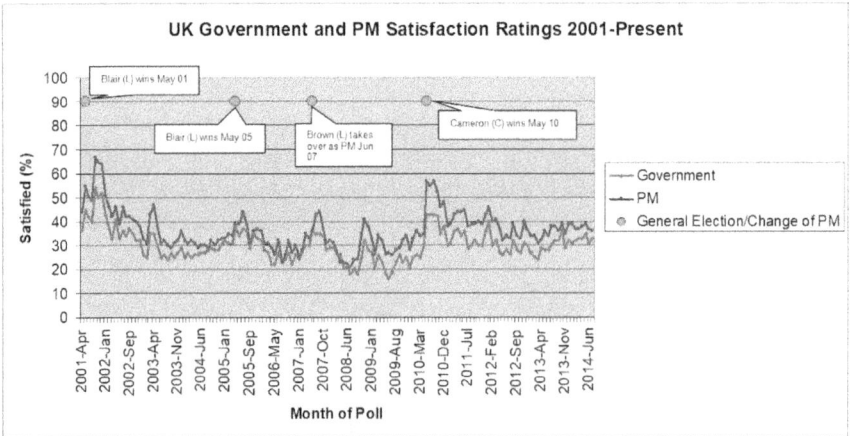

Figure 9.9   PM and Government Approval Ratings

Questions: 'Are you satisfied or dissatisfied by the way the Prime Minister is doing his job?' and 'Are you satisfied or dissatisfied with the way the Government is running the country?'

Source: Author from (IPSOS-Mori 2014) polls[48]

ratings for both the PM and the government on a continuous basis through the period covered. Tony Blair got a big bounce in September 2001 (67 per cent up from 49 per cent) after 9/11 and a smaller bounce in March 2003 (43 per cent up from 31 per cent) when the war in Iraq

---

48   'Political Monitor: Satisfaction Ratings 1997–Present' *Ipsos-Mori* (2014), available at http://www.ipsos-mori.com/researchpublications/researcharchive/88/Political-Monitor-Satisfaction-Ratings-1997Present.aspx?view=wide&view=print, accessed 25 October 2018.

began. In contrast, Gordon Brown suffered a significant dip (32 per cent to 26 per cent) at the time when the last British troops left Iraq. In the case of Libya, David Cameron got a small rise (4 per cent) in March 2011 when British warplanes were committed but, throughout the war, there was a steady monthly variance (no more than 1 per cent) until June 2011, which registered a 7 per cent drop.[49] However, this demonstrates only that foreign policy and war may have some impact on the PM's approval rating. The impact of this rating on the public's support for wars will be examined later.

Competing Models Critiqued

Having discussed the operational findings, these need to be applied as a critique of the three main models noted earlier, each of which has a derived hypothesis. The models and hypotheses are the *Ends and Means Calculus* (H1), the *Principle Policy Aim* (H2) and the *Leadership Effect Theory* (H3). The dependant variable for all three models is public support and each of the hypotheses engages with it in different ways.

Ends and Means Calculus I: The Costs

H1 (*Ends and Means*) predicted that domestic support for wars is sensitive to both the 'costs' and the public's perception of 'success'. The focus will be primarily on Iraq and Afghanistan, where Britain paid the highest costs in both human and financial terms. John Mueller argues that public support can be explained by a simple association: as casualties mount, support will decrease.[50] However, when applying this theory to Afghanistan, the bloodiest case study in the sample, the data – as noted in Table 9.1 – does not support such a clear correlation, placing the composite poll results for British public support along with the cumulative number of UK fatalities. Public support rose in July 2009, even though casualties at that point totalled 182, with 33 new deaths from February alone. This could be due to the hard fighting that was going on at the time: Operation Panther's Claw, conducted in July 2009, was widely viewed as a success, despite being costly in terms of casualties.[51]

---

49 The June 2011 drop may be partly explained by domestic factors, such as the *News of the World* phone-hacking story and public sector strikes over pensions.
50 Mueller, 'Trends in Popular Support for the Wars in Korea and Vietnam', 358–75.
51 'Afghanistan offensive "a success"', *BBC News* 27 July 2009, available at http://news.bbc.co.uk/1/hi/in_depth/8170432.stm, accessed 25 October 2018.

Date	Public Support (%)	Casualties (cumulative)
2001–October	74	0
2001–November	66	0
2006–September	31	40
2007–July	37	68
2007–August	25	74
2009–February	22	149
2009–July	46	182
2011–April	19	357
2012–May	14	416
2014–April	5	453

Table 9.1   Afghanistan: Public Support and Casualties (Cumulative)

Sources: Author from *BBC News*, *Casualty Monitor*, ICM, Ipsos-Mori, YouGov polls[52]

Date	Public Support (%)	Casualties (monthly)
2001–October	74	0
2001–November	66	0
2006–September	31	27
2007–July	37	6
2007–August	25	6
2009–February	22	6
2009–July	46	22
2011–April	19	1
2012–May	14	4
2014–April	5	5

Table 9.2   Afghanistan: Public Support and Casualties (Monthly)

Sources: Author from *BBC News*, *Casualty Monitor*, ICM, Ipsos-Mori, YouGov polls[53]

---

52   *BBC News*, 'UK Military Deaths in Iraq'; *Casualty Monitor*, 'British Casualties: Iraq'. For details of the range of Ipsos-Mori tracker polls used as part of this process, see 'Economist/ Ipsos-Mori Issues Tracker', *Ipsos-Mori*, available at https://www.ipsos.com/ipsos-mori/ en-uk/news-and-polls/overview, accessed 28 October 2018; 'YouGov/Channel 4 Iraq Commission Results on Iraq since the War', *YouGov* 5–7 June 2007, available at http:// cdn.yougov.com/today_uk_import/YG-Archives-Ira-ch4-IraqCommission-070612.pdf, accessed 28 October 2018.

53   See Note 52.

Similarly, when the poll results are broken down using month-by-month casualty figures (Table 9.2) it reinforces the cumulative findings, with public support rising to 46 per cent when H1 suggests it should be declining. The tables demonstrate that there is no clear correlation between either cumulative (overall) or month-by-month casualty figures. This is not to say that casualty figures do not matter. Rather, it suggests that Mueller's explanation may be overstated. The British public at least may be prepared to accept more casualties than is generally assumed.

Turning to Iraq, the data similarly identifies no clear correlation in trends. Table 9.3 illustrates the composite poll results for British public support along with the cumulative number of UK fatalities. There seems to be less of a casualty 'bounce', with no sudden rises in support. The initial invasion in March 2003 explains why the casualty rate was so high initially, before tapering off. The poll results, when broken down using month-by-month casualty figures (Table 9.4), do reveal a possible correlation – albeit isolated – in April 2007, where there were

Date	Public Support (%)	Casualties (cumulative)
2003–March	53	27
2003–October	46	52
2004–January	50	58
2004–October	42	70
2005–May	35	88
2005–September	35	95
2006–March	33	103
2006–July	36	115
2007–April	26	146
2007–August	19	168

Table 9.3   Iraq: Public Support and Casualties (Cumulative)
Sources: Author from *BBC News, Casualty Monitor*, ICM, Ipsos-Mori, YouGov polls[54]

16 casualties (a high monthly figure for Iraq) and public support dipped accordingly to 26 per cent from 36 per cent. As there is an absence of information about public opinion in the period from July 2006 to April 2007, the noted drop might have been more gradual over the period (indicating that the casualties were not a significant factor) or it may have been sudden.

---

54 See Note 52.

Date	Public Support (%)	Casualties (monthly)
2003–March	53	27
2003–October	46	1
2004–January	50	5
2004–October	42	2
2005–May	35	2
2005–September	35	3
2006–March	33	0
2006–July	36	2
2007–April	26	16
2007–August	19	5

Table 9.4   Iraq: Public Support and Casualties (Monthly)

Sources: Author from *BBC News*, *Casualty Monitor*, ICM, Ipsos-Mori, YouGov polls[55]

Overall, as an independent variable, casualty levels do not seem to be a reliable guide in terms of how the British public responds. Looking at longer-term trends (as noted in Figure 9.10), while the decline in support for both campaigns is steady and apparently inexorable, there is no statistical correlation between casualty levels and public support, be they cumulative

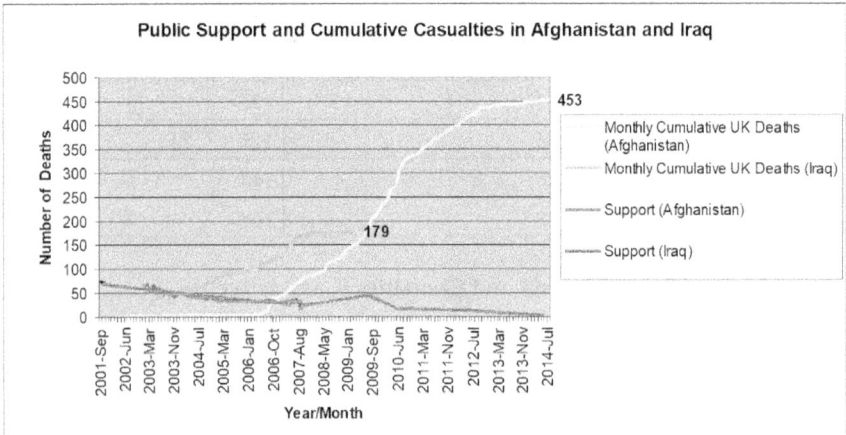

Figure 9.10   Iraq and Afghanistan: Public Support, Cumulative Casualties

Sources: Author from *BBC News*, *Casualty Monitor*, ICM, Ipsos-Mori, YouGov polls[56]

---

55 See Note 52.
56 See Note 52.

or month-by-month. Furthermore, support lines for both wars follow a very similar downward trajectory. One possibility is that duration may be a more significant factor than casualty sensitivity – an explanation that fits more with the argument that wars that fail to achieve their objectives quickly tend to lose support.[57]

Ends and Means Calculus II: The Benefits

The other side of the coin are the benefits. Gelpi, Feaver and Reifler argue that it is success, above all other factors, that matters most.[58] Sadly, the polling data is not sufficient to establish any 'benefits' patterns for either Afghanistan or Iraq. To test the 'success matters' theory the Libyan campaign is most apposite. The YouGov tracker offers a more complete measure of British opinion towards Libya,[59] with three distinct points of correlation identified. Point one denotes the initial area where the public was unsure – people were neither fully convinced nor dissuaded. Both sets of lines converge with one another ('Wells' and 'Rights'; 'Badlys' and 'Wrongs'), with public perception shifting towards the campaign going 'more badly' than 'well'. Point two, however, indicates a shift to optimism and support. From this point on the 'Wells' are in the ascendancy and the 'Rights' start to recover too – albeit more slowly. Moreover, while not an exact match, the trends mirror one another and, when the perception of success increases, levels of public support do too.

Gelpi, Feaver and Reifler's assertion that success is not measured 'in terms of body bags' seems credible. Furthermore, their model indicates that, if the public believes that victory remains achievable, it will support the war.[60] This seems to be corroborated by the Libya polling data, where success does seem to influence how the British public responds. In Figure 9.11, it is clear that a relationship exists and that the two trends are broadly similar. It would be interesting to apply the 'success' model rigorously to Iraq or Afghanistan, where the scale, intensity and duration are all significantly higher than in Libya. Despite this, much of the recent literature supports the idea of success being important. For example, Thomas Scotto uses a survey to make the case that Britain's role in Afghanistan is linked to perceptions of success.[61]

---

57 Larson, *Casualties and Consensus*.
58 See Gelpi, Feaver and Reifler, *Paying the Human Costs of War*.
59 See Clements, 'Public Opinion and Military Intervention: Afghanistan, Iraq and Libya'.
60 See Gelpi, Feaver and Reifler, 'Success Matters: Casualty Sensitivity and the War in Iraq'.
61 See Scotto, Reifler, Clarke, Lopez, Sanders, Stewart and Whiteley, *Briefing Paper: Attitudes towards British Involvement in Afghanistan*.

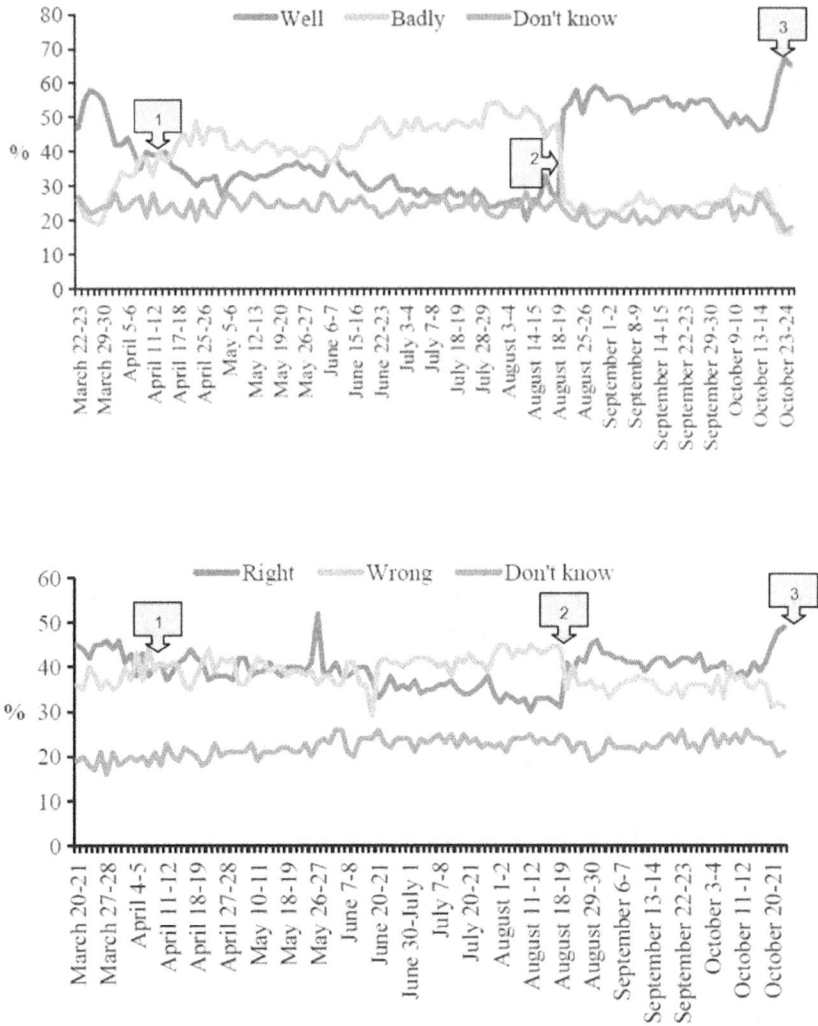

Figure 9.11   Public Perception of Success (top) and Public Support (bottom) in Libya

Source: Clements (2013) from YouGov polls[62]

In testing H1 (*Ends and Means*), one half stands up to greater scrutiny, with polling evidence suggesting that domestic support for wars is *less sensitive* to the 'costs' than is popularly thought but its perception of 'success' *is* highly significant. Indeed, the British public is not as

---

62   Clements, 'Public Opinion and Military Intervention: Afghanistan, Iraq and Libya', 128.

casualty-phobic as the MoD has suggested in the past, which could have policy implications for how politicians communicate with the electorate on matters of war.[63] Embarking upon new and potentially more complex 'wars amongst the people' could be difficult to sell unless politicians can convince the public that success remains likely.

Principle Policy Aim Model

The notion that a war can be right or wrong – or moral or immoral – is of great importance to proponents of the *Principle Policy Aim* theory.[64] Indeed, Reifler argues that the morality model's explanatory power is impressive.[65] H2 argued that domestic support for wars depended on the *Principle Policy Aim* being viewed by the public as either legitimate or in the national interest. Many would argue that the Afghanistan operation was a legitimate (even just) cause, at least initially, as the ends were framed as defeating the 9/11 perpetrators. Conversely, Iraq is popularly viewed as having been the 'bad' or 'wrong' war. However, both wars became less popular as their stated policy aims changed and their duration lengthened. Therefore, if the reasons for going to war are changed – or 'mission creep' takes place – noted levels of support should decline.

In Iraq, the publicly-stated rationale for war seemed to evolve over time, from alleged weapons of mass destruction to that of regime change and democratisation. What was initially perceived by the British public to be one kind of war morphed into another type entirely. It is worth remembering that there was a significant majority against Britain joining the US in the Iraq war without a UN resolution.[66] This may suggest some kind of public 'national interest' or even 'moral' calculation. Another study, on Afghanistan in 2009, revealed that 49 per cent of those surveyed felt the war was not morally justified.[67] This significant drop from the initially strong support post 9/11 could reflect the public's perception of changing or shifting war aims, from foreign policy restraint (persuade the Taliban

---

63 See Development, Concepts and Doctrine Centre, *Risk: The Implications of Current Attitudes to Risk for the Joint Operational Concept.*

64 See Jentleson, 'The Pretty Prudent Public', 49–74; Peter Liberman, 'An eye for an eye: Public support for war against evildoers', *International Organization* 60:3 (2006), 687–722; Rachel Stein, 'War as punishment: Retributive justice and the public support for the use of force'. Paper presented at the 2011 Annual Meeting of the Midwest Political Science Association, Chicago.

65 Reifler, Clarke, Scotto, Sanders, Stewart and Whiteley, 'Prudence, Principle and Minimal Heuristics', 39.

66 Baines and Worcester, 'When the British "Tommy" went to war, public opinion followed', 4.

67 Scotto, Reifler, Clarke, Lopez, Sanders, Stewart and Whiteley, *Briefing Paper: Attitudes towards British Involvement in Afghanistan*, 8.

to surrender Bin Laden) through to internal political change (defeat the Taliban) and finally to wider aspects of nation-building, such as women's rights and counter-narcotics. While these goals could all reasonably be considered legitimate (and moral), the public seems cautious or sceptical about principle policy aims deviating too much from those originally stated. These changing goals, if the model is sound, would certainly be an explanatory factor for the decline in support.

Additionally, and irrespective of changing aims, could the type of war being fought be a factor? The example of Syria is of value here.

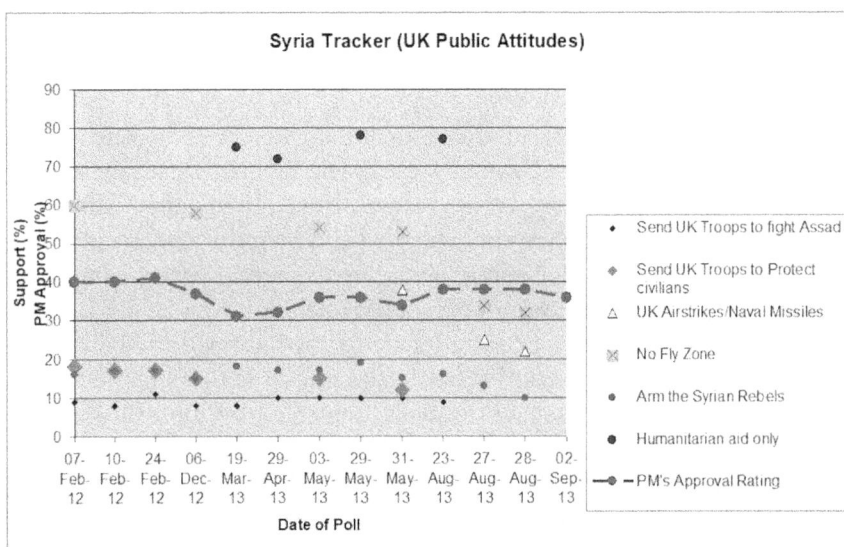

Figure 9.12   PM Approval and Syria 'Options'

Source: Author from (Ipsos-Mori and YouGov) polls[68]

Figure 9.12 represents some of the different policy options being considered in 2012–13. This is relevant because each option, while being different in scale and intensity, would have been fought for different reasons. If Jentleson's morality model is right, support for each option

---

68 'Multi-Country Euro track poll (Syria)' *YouGov* 22–27 March 2012, available at http://cdn.yougov.com/cumulus_uploads/document/ky0gqwqt2o/Cross%20country%20results%20120326%20EU.pdf, accessed 28 October 2018; for details on Ipsos-Mori and Syria, see 'Syria Pulse Survey' *Ipsos-Mori* 4 September 2013, available at https://ems.ipsos-mori.com/researchpublications/researcharchive/3250/Low-support-for-military-action-against-Syria-by-Britain-and-the-US-amid-fears-it-would-encourage-attacks-on-the-West.aspx, accessed 28 October 2018.

should be affected by the type of intervention proposed. He lists three main types of principle policy objectives: 'foreign policy restraint' (FPR), 'internal political change' (IPC), and 'humanitarian intervention' (HI).[69] Indeed, what is striking about the Syria data is just how consistent support patterns are for each 'objective'. Moreover, while support for each objective rises or falls, at only one point does the order of popular support change: initially support for sending 'UK troops to protect civilians' was higher than 'arm the Syrian rebels'. Apart from this 'blip', the order of public support for the objectives remains constant: (1) Humanitarian aid only (HI); (2) No fly zone (FPR); (3) UK airstrikes or use of at-sea naval missiles (FPR); (4) Arm the Syrian rebels (IPC); (5) Use ground troops to protect civilians (HI) and (6) Use ground troops to fight Assad (IPC).

The model predicts that the order should unfold by the group or type of objective but it does not. Instead, the order seems more to reflect the perceived riskiness of the choice. This seems more in line with the *Ends and Means Calculus* and *Cost-Benefit* models (H1). Policymakers should think seriously about the selection and maintenance of the aim, as the reflexive response that 'something must be done' is not good enough. If they fail to do this, they are likely to lose public support. Moreover, changing principle policy aims is also potentially damaging.

The Leadership Effect Model

Predictors in the *Leadership Effect* model suggest that popular leaders may be better able to convince voters that a war is worth fighting,[70] with domestic support for wars influenced by a leader who has a high public approval rating. To test this, it is instructive to comprehensively examine PM approval ratings (Figure 9.13). There are several trends that should be observed to prevent any misreading or misinterpretation of the data. First, the figure shows that all three PMs received a boost after coming to power, regardless of the means of doing so. In effect, novelty provides its own boost. Second, for the three examples considered, all proved more popular than the government they led, most of the time. This could be because British PMs are not blamed as much by the public for the government's day-to-day performance or it could be that the individuals were particularly

69   Jentleson, 'The Pretty Prudent Public', 49–74.
70   See Baines and Worcester, 'When the British "Tommy" went to war, public opinion followed'; Baker and O'Neal, 'Patriotism or opinion leadership', 661–87; Baum and Potter, 'The Relationships between Mass Media, Public Opinion and Foreign Policy', 39–65; Karbo, 'Prime Minister Leadership Styles in Foreign Policy Decision making', 553–81; Zaller, *The Nature and Origins of Mass Opinion*.

charismatic, enjoying a significant 'halo' effect.[71] However, any 'halo' effect would assumedly be less pronounced or even potentially negative for Brown, widely regarded as the least charismatic of the three (as is validated by the polling statistics). The final possibility is that government approval might instead be closely correlated to PM approval. This would mean that, instead of two independent variables, the PM approval variable would be the dependant one; this seems, from the available data, to be the most likely explanation. This does not invalidate the possibility of a 'halo' effect, but it does mean it would be relative between different PMs and not determined by government approval (or lack thereof).

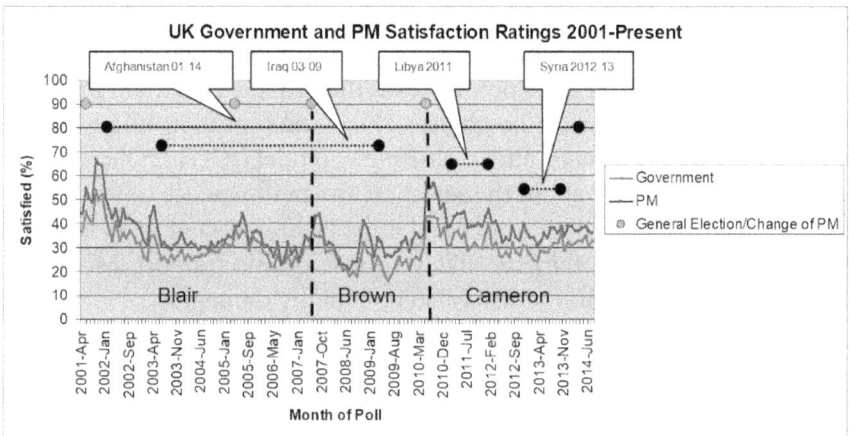

Figure 9.13   PM Approval and Dates of Wars and PM Tenures

Source: Author from (ICM, Ipsos-Mori, YouGov) polls[72]

Taking Syria as a useful example, the prospect of British intervention was deeply unpopular from the beginning. As noted above, only the provision of humanitarian aid was palatable to the public. The surprising thing about Cameron's approval ratings at this time is just how steady they are (Figure 9.14). They neither affected nor were affected by the public mood on Syria. Several explanations are possible. One theory is that the public did not care enough about Syria, with his ratings primarily reflecting domestic concerns. Implicit in this is that, when the public cares more, ratings would

---

71 Dean, 'The Halo Effect: When Your Own Mind is a Mystery'; Murphy, Jako and Anhalt, 'Nature and consequences of halo error', 218–25.

72 'Political Monitor: Satisfaction Ratings 1997–Present' *Ipsos-Mori* 19 January 2017, available at http://www.ipsos-mori.com/researchpublications/researcharchive/88/Political-Monitor-Satisfaction-Ratings-1997Present.aspx?view=wide&view=print, accessed 28 October 2018.

UK Government and PM Satisfaction Ratings During Syria Crisis

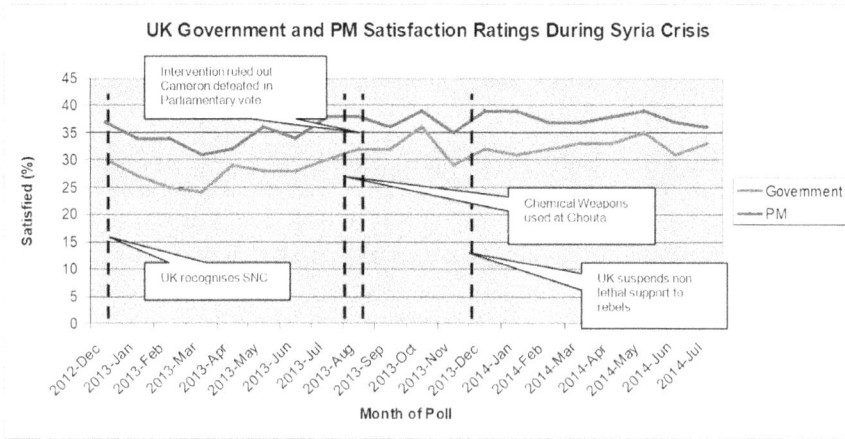

Figure 9.14   PM and Government Approval Ratings (Syria)

Source: Author from (IPSOS-Mori 2014) polls[73]

alter accordingly. Another possibility is that the leadership theory has less of an effect than the literature suggests, with Reifler noting that 'the impacts of various heuristics are much weaker' with regard to the British public.[74]

Perhaps the *Leadership Effect* is most relevant when initially convincing the public to go to war. Figure 9.15 records approval levels at the beginning and end of each conflict. Significantly, Cameron's approval rating was one

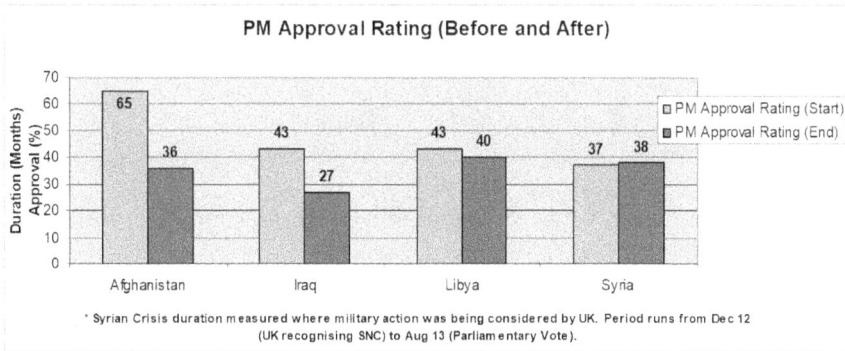

PM Approval Rating (Before and After)

* Syrian Crisis duration measured where military action was being considered by UK. Period runs from Dec 12 (UK recognising SNC) to Aug 13 (Parliamentary Vote).

Figure 9.15   PM Approval Ratings (Before and After)

Sources: Author from (IPSOS-Mori) polls[75]

---

73 Ipsos-Mori, 'Political Monitor: Satisfaction Ratings 1997–Present'.
74 Reifler, Clarke, Scotto, Sanders, Stewart and Whiteley, 'Prudence, Principle and Minimal Heuristics', 49.
75 Ipsos-Mori, 'Political Monitor: Satisfaction Ratings 1997–Present'.

point higher after being prevented from undertaking a military inter-vention in Syria (37 per cent to 38 per cent). Blair rode an unprecedented wave of popularity (65 per cent) when he committed Britain to war in Afghanistan. However, the post 9/11 response both in the UK and the US has potentially more to do with the *Principle Policy Aim* model and H2. Blair's second 'bounce' in September 2001 was almost certainly due to public feeling about 9/11.

With Iraq, Blair received a post-invasion 'bounce' that quickly disappeared as the war moved into its stabilisation and counterinsurgency phase. This is consistent with – and therefore seems to validate more – the 'rally-round-the-flag' notion, whereby the public supports the troops regardless of their views on either the legitimacy or benefits of the war.[76] Equally, it could be linked to perceptions of success – as in H1 – as the initial fighting against Saddam Hussein's army was both quick and comprehensive. The 'before' and 'after' ratings are more pronounced for both Afghanistan (a 31-point differential) and Iraq (a 16-point differential). Libya and Syria are relatively stable, shifting only 3 per cent and 1 per cent respectively. However, caution should be exercised, not least because the longer examples spanned multiple leaders, while the figures for Libya and Syria all relate to David Cameron. This suggests that, while politicians may not be able to shape public opinion in the longer term, they do benefit (albeit briefly) when kinetic operations begin. The initial 'rally-round-the-flag' effect should not be overstated; nor can it validate longer-term support. Leaders should therefore endeavour to keep wars short wherever possible, although the increased complexity of 'war amongst the people' may run contrary to that.

Conclusion

Scholarship remains divided over which factor is most important in determining British public support for war. By using data gathered from several polling organisations over a 13-year period, spanning three wars and one crisis, this chapter has presented a more differentiated picture of British public attitudes. The case studies were used as a prism to view different models and were tested against three hypotheses, producing a number of significant results.

First, H1 was partly proven in that public perception of success matters much more than the costs. This validates the argument that success

---

76 Baker and O'Neal, 'Patriotism or opinion leadership', 661–87.

matters most.[77] While costs probably do form part of the overall *Means and Ends Calculus*, the findings assessed here indicate that Mueller's model[78] is not as readily applicable, with conflict duration having better explanatory power. The British public is not as casualty averse as assumed and *are* prepared to pay the costs, at least initially. Furthermore, it suggests that cost-benefit analysis goes on with the public, which may well be linked to H2.

H2 was also proven: legitimacy and the reasons for entering a war matter. The *Principle Policy Aim* model suggests that *why* a war is fought is important to the public. Jentleson's model did not predict the results expected in Syria – indeed the degree of riskiness of the operation seemed more significant than the type of operation. However, the notion that legitimacy counts was borne out by the Iraq War case study; an interesting corollary was that shifting or changing aims tended to reduce public support.

Third, H3 was not proven. Elite cues, as measured by leader approval, had a minimal effect on the British public. No significant evidence was found that a popular leader can influence the public more than an unpopular one, with no significant 'halo' effect evident from the polling analysis. What is more, while wars do have a short-term effect on leader approval ratings, this dissipates quickly. 'Rally-round-the-flag' theory[79] explains this well and is the most likely reason for the pattern found. It suggests that PM approval is shaped more by the initial stages of a war than vice versa. Taken together, these findings strongly suggest that a combination of the *Principle Policy Aim* and perceived success were the most important factors in determining British public opinion to war. Casualties and the effect of leaders mattered much less.

Moving from theory to policy, the implications are clear. The British public is not casualty-phobic; policymakers can be assured that this is not the primary driver of public opinion towards war. Instead officials should seek to convince the public that future wars – of whatever type, however complex – are 'winnable'. Once committed to war, the strategic messaging and communications plans must be used to reinforce operational successes. Libya proves that the perception of operational success can bolster public support for a campaign. In terms of duration, leaders should try to keep conflicts short. The 'rally-round-the-flag' effect gives an immediate boost once the fighting starts but public support quickly returns to 'normal' levels. Additionally, politicians must select clearly defined and legitimate

77   Gelpi, Feaver and Reifler, *Paying the Human Costs of War*.
78   Mueller, 'Trends in Popular Support for the Wars in Korea and Vietnam', 358–75.
79   Baum and Potter, 'The Relationships between Mass Media, Public Opinion and Foreign Policy', 39–65.

policy aims and stick to them. Iraq and Afghanistan demonstrate that the public is not tolerant of politicians who take them to war for one reason and then subsequently change it to another. People seem to make a rational assessment based on the likelihood of success balanced against the legitimacy of the aim. Understanding this calculus could enable policymakers and military leaders to make more informed and better choices. In the twenty-first century, debates surrounding the concept of 'war amongst the people' should be expanded to include the intervening state's public and refocused to better understand how their views are shaped and formed. Policymakers and generals who fail to win 'hearts and minds' on the *home* front otherwise risk falling into Bill Rammel's 'inglorious and impotent isolation by default'.

References

Baines, P. and Worcester, R., 'When the British "Tommy" went to war, public opinion followed', *Journal of Public Affairs* 5 (2005).

Baker, W. and O'Neal, J., 'Patriotism or opinion leadership? The nature and origins of the "rally round the flag" effect', *Journal of Conflict Resolution* 44:5 (2001).

Baum, M., 'The constituent foundations of the rally-round-the-flag phenomenon', *International Studies Quarterly* 46:2 (2002).

Baum, M. and Potter, P., 'The Relationships Between Mass Media, Public Opinion and Foreign Policy: Towards a Theoretical Synthesis', *Annual Review of Political Science* 11 (2008).

Berinsky, A. and Druckman, J., 'Public Opinion research and support for the Iraq War', *Public Opinion Quarterly* 71:1 (2007).

Berinsky, A. and Kinder, D., 'Making Sense of Issues through Media Frames: Understanding the Kosovo Crisis' *The Journal of Politics* 68:3 (2006).

Canan-Sokullu, E., 'Domestic Support for Wars: A Cross-Case and Cross-Country Analysis', *Armed Forces and Society* 38:38 (2011).

Clements, B., 'Public Opinion and Military Intervention: Afghanistan, Iraq and Libya', *The Political Quarterly* 84:1 (2013).

Dean, J., 'The Halo Effect: When Your Own Mind is a Mystery', *PsyBlog* 31 October 2007, available at http://www.spring.org.uk/2007/10/halo-effect-when-your-own-mind-is.php, accessed 25 October 2018.

Development, Concepts and Doctrine Centre, *Risk: The Implications of Current Attitudes to Risk for the Joint Operational Concept* (Shrivenham: Ministry of Defence, 2012).

Gelpi, C., Feaver, P. and Reifler, J., 'Success Matters: Casualty Sensitivity and the War in Iraq', *International Security* 30:3 (2006).

Gelpi, C., Feaver, P. and Reifler, J., *Paying the Human Costs of War: American Public Opinion and Casualties in Military Conflicts* (New Jersey: Princeton University Press, 2009).

Hermann, R., Tetlock, P. and Visser, P., 'Mass public decisions to go to war: A cognitive-interactionist approach', *American Political Science Review* 93:3 (1999).

Jacobsen, G., 'A Tale of Two Wars: Public Opinion on the US Military Interventions in Afghanistan and Iraq', *Presidential Studies Quarterly* 40:4 (2010).

Jentleson, B., 'The Pretty Prudent Public: Post-Vietnam American Opinion on the Use of Military Force', *International Studies Quarterly* 36:1 (1992).

Karbo, J., 'Prime Minister Leadership Styles in Foreign Policy Decision-making: A Framework for Research', *Political Psychology* 18:3 (1997).

Kellner, P., 'Syria and the long shadow of Iraq', *Daily Telegraph* 28 August 2013 available at https://www.telegraph.co.uk/news/worldnews/middleeast/syria/10270344/Peter-Kellner-Syria-and-the-long-shadow-of-Iraq.html accessed 25 October 2018.

Kuklinski, J. and Quirk, P., 'Reconsidering the rational public: Cognition, heuristics and mass opinion' in Arthur Lupia, Matthew McCubbins and Samuel Popkin (eds), *Elements of Reason* (Cambridge: Cambridge University Press, 2000).

Kull, S., 'Review of Eric Larsen's *Casualties and Consensus*', *Public Opinion Quarterly* 61 (1997).

Larson, E., *Casualties and Consensus: The Historical Role of Casualties in Domestic Support for US Military Operations*, (Santa Monica: RAND, 1996).

Liberman, P., 'An eye for an eye: Public support for war against evildoers', *International Organization* 60:3 (2006).

Mueller, J., 'Trends in Popular Support for the Wars in Korea and Vietnam', *American Political Science Review* 65:2 (1971).

Murphy, K., Jako, R., and Anhalt, R., 'Nature and consequences of halo error: A critical analysis', *Journal of Applied Psychology* 78:2 (1993).

Quinn, B., 'MoD study sets out how to sell wars to the public', *The Guardian* 26 September 2013, available at https://www.theguardian.com/uk-news/2013/sep/26/mod-study-sell-wars-public, accessed 25 October 2018.

Reifler, J., Clarke, H., Scotto, T., Sanders, D., Stewart, M. and Whiteley, P., 'Prudence, Principle and Minimal Heuristics: British Public Opinion toward the Use of Military Force in Afghanistan and Libya', *British Public Opinion* 16:1 (2014).

Scotto, T., Reifler, J., Clarke, H., Lopez, J., Sanders, D., Stewart, M. and Whiteley, P., *Briefing Paper: Attitudes towards British Involvement in Afghanistan* (Essex: Institute for Democracy and Conflict Resolution 2011).

Stein, R., 'War as punishment: Retributive justice and the public support for the use of force'. Paper presented at the 2011 Annual Meeting of the Midwest Political Science Association, Chicago.

Zaller, J., *The Nature and Origins of Mass Opinion* (Cambridge: Cambridge University Press, 1992).

10

'WAR AMONGST THE PEOPLE'

Responses in British Land Tactical Doctrine

John Bailey

Introduction

According to General Sir Rupert Smith's 'war amongst the people' thesis, recent conflicts have exposed a 'paradigm shift' in the execution of war.[1] Increasingly, industrial interstate warfare is being replaced by the deployment of states' forces into more complex environments for the conduct of operations 'amongst the people'.[2] These environments contain audiences, actors, adversaries and enemies with varying motives, loyalties, capabilities and propensities for violence. Rather than attempting to defeat enemies by trial of strength to settle international disputes, Smith proposes that the objective of military intervention is to change the intentions, or capture the will, of all these groups, creating 'a condition in which a strategic result is achieved'.[3] In this setting of shifting thinking on the changing character of war, the British Army's doctrine of stability operations has

---

1   This chapter examines the treatment, within land tactical doctrine, of the themes contained within General Sir Rupert Smith's 2005 'war amongst the people' thesis. While not all of the ideas in the thesis are exclusive to him, and the exact extent of his influence on doctrine writers cannot be measured, the chapter employs his work to illustrate how recent discourse concerning the changing character of war has impacted upon doctrine. See Rupert Smith, *The Utility of Force* (London: Allen Lane, 2005), 3.

2   Smith, *The Utility of Force*, 267.

3   Toni Pfanner, 'Interview with General Sir Rupert Smith', *International Review of the Red Cross* 88:864 (2006), 719–27.

evolved, complementing the more traditional ways of warfare through which land forces can support the ends of national security policy.[4]

This chapter explains how, over the last five years, authors of such doctrine have adapted their publications to the characteristics of contemporary conflict, as described by Smith and other commentators. At the same time, in a sign that Smith's views have not been accepted wholesale, it will demonstrate that doctrine writers continue to allow for the possibility of industrial interstate war.

The chapter begins by situating the author's views.[5] Thereafter, it places Smith's 'war amongst the people' thesis in the context of discourse on the meaning of security and the changing character of war and explains how doctrine is created. It continues by identifying influences on doctrine design, listing and describing current doctrine before discussing factors affecting soldiers' use of doctrine for 'war amongst the people'. Finally, the chapter concludes by speculating on the future relevance of Smith's thesis to doctrine writers.

Analytical Lens

This chapter is written by a serving army officer closely involved in the production of land tactical doctrine.[6] The analysis is based on personal experience, academic discourse, the themes presented within official doctrine publications and the Government's policy documents. Given his affiliation, he is at risk of privileging the British State as *the* referent security object within the contemporary operating environment.[7] Nonetheless, the chapter will demonstrate that other voices promoting non-traditional,

---

4   Stability operations describe the military contribution to stabilisation. Stabilisation 'is one of the approaches used in situations of violent conflict which is designed to protect and promote legitimate political authority, using a combination of integrated civilian and military actions to reduce violence, re-establish security and prepare for longer-term recovery by building an enabling environment for structural stability'. See Stabilisation Unit, *The UK Government's Approach to Stabilisation* (London: Stabilisation Unit, 2014), available at http://sclr.stabilisationunit.gov.uk/publications/stabilisation-series/487-uk-approach-to-stabilisation-2014/file, accessed 1 November 2018.

5   While elements of General Smith's thesis are criticised here, the author recognises not only his impressive military record but also his contribution to debates concerning the changing character of war. Indeed, it is mainly as a consequence of the passage of time and through subsequent events that fresh perspectives have formed on his thesis.

6   The author seeks to promote the concept of insider-ness, namely the practice of insiders commenting on and analysing organisations of which they are part. See David Walker, 'Putting "Insider-ness" to Work' in Alison Williams, Neil Jenkins, Matthew F. Rech and Rachel Woodward (eds), *The Routledge Companion to Military Research Methods* (Abingdon: Routledge, 2016), 256–67.

7   The referent security object describes the 'thing' that is to be protected.

'population-focused' security outlooks are acknowledged in doctrine and that the views of non-military experts involved in stabilisation are also an important element in its production.[8] These views are considered here because they align with elements of Smith's thesis – specifically, his view that 'in our modern conflicts, dealing with the civilian population is directly associated with the objective and is a primary not secondary activity'.[9] In embracing such alternative and external viewpoints, referent security objects beyond the state can be, and are, recognised within doctrine. These alternative views must be heard within the military institution because recent Government strategy papers reflect the same 'deepened and broadened' interpretations of security as have appeared in academia.[10] Nonetheless, it is important to recognise that for some commentators, such as Kienscherf and Suhrke, the promotion of non-traditional security approaches by states has the power to mask their 'power-oriented interests'.[11]

The Meaning of Security, 'War Amongst the People' and the Changing Character of War

Smith's thesis has salience in contemporary debates concerning the meaning of security and the changing character of war. The link between these two themes is that the establishment of referent security objects other than the state has the potential to alter the character of the missions and mandates for which British land forces are prepared. This means that land tactical doctrine has had to be significantly adapted to encompass guidance for its forces involved in operations beyond conventional interstate warfare.

Militaries and the 'Deepening and Broadening' of the Meaning of Security

Since the end of the Cold War, the meaning of security has been 'deepening and broadening' in both academic and political circles.[12] Explicitly, beyond

---

8   See David Mutimer, 'Critical Security Studies: A Schismatic History' in Alan Collins (ed.), *Contemporary Security Studies* (Oxford: Oxford University Press, 2016), 87–107; Columba Peoples and Nick Vaughan-Williams, *Critical Security Studies* (London: Routledge, 2015), 1–13.

9   Smith, *The Utility of Force*, 387–88.

10   Alan Collins, 'What is Security Studies?' in Collins (ed.), *Contemporary Security Studies*, 6.

11   For example, states' neo-realist ideologies promoting human rights, international humanitarian law and 'equitous' socio-economic development have been present in narratives surrounding foreign interventions. See Markus Kienscherf, 'Producing "responsible" self-governance: counterinsurgency and the violence of neo-liberal rule', *Critical Military Studies* 2:3 (2016), 173–92; Astri Suhrke, 'Human Security and the Interests of States', *Security Dialogue* 30:3 (1999), 265–66.

12   Collins, 'What is Security Studies?', 6.

Weberian preoccupations with the security of states and their territories, other schools of thought have emerged calling for consideration of alternative security objects. The most prominent of these are the Copenhagen and Aberystwyth Schools. Their positions are worthy of note because traces of these approaches have appeared within government and UN policy, both of which influence land tactical doctrine – the focus of this chapter.[13]

The Copenhagen School is characterised by a sectoral approach to security and the idea of securitisation. The sectoral approach considers security from, amongst others, political, societal, economic and environmental perspectives, rather than being focused on a state's territorial integrity alone. This is not to suggest, however, that these sectors do not contribute to the security of a state as a whole. Instead, the sectoral approach allows for a more nuanced analysis of its component parts. The idea of securitisation allows any actor to construct a security object, threatened existentially, which must be protected.[14] Through the act of securitisation 'a concern is framed as a security issue and moved from the politicized to the securitized'.[15] Of note, these ideas are reflected in two recent land tactical doctrine publications: *Doctrine Note 16/02: Human Security: The Military Contribution* and *Army Field Manual: Tactics for Stability Operations*, which consider a range of security objects and the impact of conflict on multiple actors.[16] The former considers the role of land forces in protecting civilians from sexual violence within peace support operations, a security concern seemingly removed from the issue of the UK's territorial integrity. The Aberystwyth School, associated with the Critical Security Studies movement, also challenges the state-centric paradigm but prioritises human emancipation above the security objects proposed by the Copenhagen School. These ideas can be seen in UN doctrines of Responsibility to Protect (R2P) and the Protection of Civilians, both of which have been subsumed, in part, within land tactical doctrine, as will be discussed below.

The ideas contained within the above schools also chime with elements of Smith's thesis. He has argued, for instance, that to 'succeed'

---

13 For example, HM Government, *UK National Action Plan on Women, Peace and Security 2018–22* (London: HM Government, 2018), available at https://www.gov.uk/government/uploads/system/uploads/attachment_data/file/677586/FCO1215-NAP-Women-Peace-Security-ONLINE_V2.pdf, accessed 1 November 2018; United Nations Trust Fund for Human Security, *Human Security* (New York: United Nations, 2018), available at https://www.un.org/humansecurity/, accessed 1 November 2018.

14 Barry Buzan, Ole Wæver and Jaap de Wilde, *Security: A New Framework for Analysis* (Boulder: Lynne Rienner, 1998), 3.

15 Collins, 'What is Security Studies?', 449.

16 UK Warfare Branch, *Human Security: The Military Contribution* (Warminster: Warfare Branch, 2016); UK Warfare Branch, *Army Field Manual: Tactics for Stability Operations* (Warminster: Warfare Branch, 2017).

within contemporary operating environments, intervening forces must capture the will of the opponent and the people amongst which they operate. In a peace support context, this means providing a credible threat of force and, as a consequence, being able to protect the civilian population. In this regard, Smith's frustration at being unable to protect civilians while commander of the United Nations Protection Force (UNPROFOR) in the face of Bosnian Serb aggression is clearly apparent within his thesis.[17] Simply put, his force lacked the utility to contend with the nature of his mandate, centred on the protection of civilians.

Nonetheless, while the Army's experiences of peace support in Bosnia in the 1990s, and more recently in South Sudan and Somalia, have allowed non-traditional security theories to influence doctrine, the security of the state remains the primary focus for land forces. In this vein, expectations that the Army will shift wholesale from its traditional warfighting remit to a more 'cosmopolitan' role are misplaced. Explicitly, Gilmore's view that Western armies might become part of some 'Global Community Policing' movement at the expense of the protection of their own states seems especially optimistic.[18] Furthermore, for some, the utility of military forces beyond combat is strictly limited. While describing the role of the American military, former Vice President Joseph Biden's views on the problematic nature of its non-warfighting remit resonate within the British context. Specifically, he notes that the dominance of the military in foreign policy may inadvertently limit options surrounding intervention, skew the balance between economic and military aid and displace other development actors.[19] Accordingly, it is unsurprising that the British Army's current capstone doctrine publication, *Army Doctrine Publication Land Operations*, states that 'the primary purpose of land forces is to conduct combat operations – to apply or threaten force'.[20]

The Changing Character of War

This book privileges Smith's 'war amongst the people' thesis as a description of the contemporary operating environment. It is by no means certain, however, that it holds a monopoly of influence on doctrine writers, given

---

17 Smith, *The Utility of Force*, 332–70.
18 Jonathan Gilmore, *The Cosmopolitan Military: Armed Forces and Human Security in the 21st Century* (London: Springer, 2015), 1–12.
19 See Joseph Biden, *Defining the Military's Role towards Foreign Policy: Hearing before the Committee on Foreign Relations on 31 July 2008* (Washington DC: United States Senate, 2008), available at https://www.gpo.gov/fdsys/pkg/CHRG-110shrg48042/pdf/CHRG-110shrg48042.pdf, accessed 1 November 2018.
20 UK Warfare Branch, *Army Doctrine Publication Land Operations* (Warminster: Warfare Branch, 2016), 8C-2.

the present broad discourse on the changing character of war. Equally, it must be conceded that Smith's work is more than a decade old and was written in the specific context of his involvement in the Balkans conflict and against the backdrop of the unfolding execution of the GWOT. In this regard, Smith has been labelled a 'new war' theorist, being placed alongside thinkers such as Lind, Van Creveld and Kaldor.[21] These theorists hold the common view that the rise of non-state actors in armed conflict, and the apparent lack of utility of state forces in defeating them, are indicative of a significant change in the character of war, if not its nature. Smith extends the challenge posed by non-state actors by identifying that conventional forces are compelled to contend with them in constrained operating environments 'amongst the people'. Nonetheless, critics have challenged the novelty of such a threat. For example, Ferguson and Roberts suggest that warfare has, to a degree, always been conducted 'amongst the people'.[22] In detail, Ferguson observes that 'war amongst the people' long predates 'the era of decolonization and superpower proxy wars' that Smith identifies as the start of the phenomenon.[23] Equally, Smith is not the first to describe the problems of conducting operations amongst civilian populations, with Kitson, Mockaitis, Nagl and Thompson all having contributed material in this field.[24] Further, White has estimated that, over the millennia, 85% of the victims of warfare have been non-combatants.[25] This estimate is supported by Clodfelter, who charts the impact of war on both militaries and civilians since the sixteenth century.[26] Additionally, for some 'new war' theorists,

---

21 This is described as such in Colin Fleming, 'New or Old Wars? Debating a Clausewitzian Future', *Journal of Strategic Studies* 32:2, 232. See also William Lind, Keith Nightengale, John Schmitt, Joseph Sutton and Gary Wilson, 'The Changing Face of War: Into the Fourth Generation', *Marine Corps Gazette* (October 1989), 22–26; Martin van Creveld, *The Transformation of War* (New York: The Free Press. 1991); Mary Kaldor, *New and Old Wars* (Cambridge: Polity Press, 2001).

22 See Niall Ferguson, 'Ameliorate, Contain, Coerce, Destroy', *The New York Times* 4 February 2007, available at http://www.nytimes.com/2007/02/04/books/review/Ferguson.t.html?pagewanted=all, accessed 1 November 2018; Sir Adam Roberts, 'The Utility of Force by Rupert Smith', *The Independent* 11 November 2005, available at https://www.independent.co.uk/arts-entertainment/books/reviews/the-utility-of-force-by-rupert-smith-326177.html, accessed 1 November 2018.

23 Ferguson, 'Ameliorate, Contain, Coerce, Destroy'.

24 See General Sir Frank Kitson, *Low Intensity Operations: Subversion, Insurgency and Peacekeeping* (London: Faber & Faber, 1971); Tom Mockaitis, *British Counter-Insurgency, 1919–1960* (Basingstoke: Macmillan, 1990); John Nagl, *Learning to Eat Soup with a Knife: Counterinsurgency Lessons from Malaya and Vietnam* (Westport: Praeger, 2002); Sir Robert Thompson, *Defeating Communist Insurgency* (London: Chatto & Windus, 1967).

25 Matthew White, *Atrocities: The 100 Deadliest Episodes in Human History* (New York: Norton, 2012), 554.

26 Michael Clodfelter, *Warfare and Armed Conflicts: A Statistical Encyclopedia of Casualty and Other Figures, 1492–2015* (Jefferson: McFarland, 2017), 4–7.

political violence is increasingly explained by identity politics, as opposed to state interest, calling into question the Clausewitzian model that views war as an extension of politics by other means. Nonetheless, Bassford, Strachan and Scheipers identify that Clausewitz's trinity does not assume that political violence is the sole preserve of states, arguing that it can be perpetrated by non-state actors.[27]

At the same time, increased tensions between the US, its allies and Russia and China have reminded observers of the possibility of interstate war, albeit unlikely given the prospect of Mutually Assured Destruction (MAD) and the suicidal outcomes for those states.[28] Of note, Cohen cites the bellicose possibilities emanating from mounting tensions in the Far East as a rebuff to Smith's view that 'battle in a field between men and machinery … as a massive deciding event in a dispute in international affairs … no longer exists'.[29]

While 'new war' theory may have its critics, some of its characteristics continue to resonate within military discourse. In defence of Smith's thesis, many of his ideas subsequently materialised within the concept of 'persistent conflict', featuring within recent British and American military publications.[30] Proposed by US General George S. Casey, it describes the present as being marked by 'protracted confrontation among state, non-state and individual actors who will use violence to achieve, political, religious, and other ideological ends'.[31] Arguably, this notion of 'persistent conflict' brings together many of the themes discussed within Security Studies, amidst the uncertainty of the immediate post-Cold War period. Crucially, aligning with Smith, it supports the view that military forces must be adapted and trained for operations beyond conventional warfighting.

---

27 See Christopher Bassford, 'Tip-Toe Through the Trinity or The Strange Persistence of Trinitarian Warfare', *Clauewitz.com* 14 April 2018, available at http://www.clausewitz.com/mobile/trinity8.htm#Non, accessed 1 November 2018; Sir Hew Strachan and Sibylle Scheipers, 'Introduction: The Changing Character of War' in Hew Strachan and Sibylle Scheipers (eds), *The Changing Character of War* (Oxford: Oxford University Press, 2011), 3.

28 Eliot Cohen, 'The End of War as We Know It', *The Washington Post* 21 January 2007, available at http://www.washingtonpost.com/wp-dyn/content/article/2007/01/18/AR2007011801981.html, accessed 1 November 2018; Colin Gray, 'Future Warfare or the Triumph of History', *Royal United Services Institute Journal* 150:5 (October 2005), 16–19.

29 Smith, *The Utility of Force*, 1.

30 UK Warfare Branch, *Army Doctrine Publication Land Operations* (Warminster: Headquarters Field Army, 2017), 1–3, 1–19, available at https://www.gov.uk/government/uploads/system/uploads/attachment_data/file/605298/Army_Field_Manual__AFM__A5_Master_ADP_Interactive_Gov_Web.pdf, accessed 1 November 2018; Department of the Army, *Operations: Field Manual 3-0; Headquarters, Department of the Army* (Washington DC: DIANE, 2008), 1-1.

31 General George Casey and Pete Geren, *A statement on the posture of the United States Army 2008 on the 26 February 2008* (Washington DC: US Government, 2008), 2, available at https://www.army.mil/aps/08/APS2008.pdf, accessed 1 November 2018.

The protracted element in 'persistent conflict' links neatly with Smith's view that 'our conflicts tend to be timeless'.[32] Further, General Sir Nicholas Carter, then Chief of the General Staff (CGS), also championed persistent conflict's corollary, the doctrine of 'persistent engagement'. This latter concept describes the approaches required to reduce the former's effects.[33] Carter's promotion of 'persistent engagement' can be seen within his *Army 2020* programme and a range of speeches made at the Royal United Services Institute (RUSI) and Chatham House.[34] Indeed, his leadership on this matter led to the 1st (UK) Division, at a scale unseen previously, being assigned the task of defence engagement in order to deliver 'persistent engagement'. At the tactical level, this has involved supporting the ends of national security through security capacity building, interoperability and security cooperation. Within the doctrine of 'persistent engagement', defence engagement activity also provides early warning of threats likely to lead to major conflict. The link between the doctrine of 'persistent engagement' and Smith's thesis is most clear in his suggestion that 'we are now in a world of continual confrontation and conflicts in which the military endeavour to support the achievement of the desired outcome by other means'.[35]

Creating Doctrine for 'War Amongst the People'

Military doctrines are a critical element within any state's security policy. They set out how military ways and means should be employed to support national grand strategic ends.[36] Accordingly, to maximise strategic benefits, alignment between ends, ways and means is essential. Yet, given the dynamic character of the contemporary operating environment and, as some have argued, the possibility that the very nature of war is changing,

---

32 Smith, *The Utility of Force*, 369.
33 Colonel Chadwick Clark and Lieutenant Colonel Richard Kiper, 'Strategic Thinking in an Era of Persistent Conflict', *Military Review* (May–June 2012), 25–34.
34 British Army, *Modernising to face an unpredictable future: Transforming The British Army* (Andover: British Army, 2012), available at https://web.archive.org/web/20130418031611/http://www.army.mod.uk/documents/general/Army2020_brochure.pdf, accessed 1 November 2018; General Sir Nicholas Carter, *Keynote speech by the Chief of the General Staff at the RUSI Land Warfare Conference on 30 June 2015* (London: RUSI, 2015), available at https://rusi.org/sites/default/files/20150630-lwc15-transcripts-cgs_opening_keynote.pdf, accessed 1 November 2018; General Sir Nicholas Carter, *Opening address by the Chief of the General Staff at the RUSI Land Warfare Conference on 28 June 2016* (London: RUSI, 2016), available at https://rusi.org/sites/default/files/160628-lwc16-cgs-opening_address.pdf, accessed 1 November 2018. The speech can also be viewed at https://www.youtube.com/watch?v=K_MRagWu8p4, accessed 1 November 2018.
35 Pfanner, 'Interview with General Sir Rupert Smith', 720.
36 Barry Posen, *The Sources of Military Doctrine: France, Britain and Germany between the World Wars* (London: Cornell University Press, 1984), 13–33.

this demand is challenging because doctrine can only represent a snapshot of thinking in time. Further, while academic debate rages regarding the changing character of war, the doctrine writer must bridge the gap between such debates, government policy and the pressing operational needs of practitioners.[37] Consequently, doctrine is often a manifestation of compromise, rarely satisfying all parties.

In the land environment – the focus of this chapter – the organisation charged with tackling this difficult task is the Warfare Branch, part of Headquarters Field Army. Recognising that the contemporary operating environment demands the provision of doctrine for both combat and stability operations, its teams are organised to support both efforts. Doctrine may be generated or updated for a variety of reasons, usually following 'top-down' pressure to innovate or 'bottom-up' insights suggesting adaptations are required.[38] These 'bottom-up' insights are usually provided through the Army's lesson learning processes, including post-operational reports and interviews. Drafts are then circulated throughout land forces and beyond for comment, before the final version is sanctioned either by the head of the Branch or, in the case of key documents, by a senior group known as the Land Environment Doctrine Committee.

Influences on Land Tactical Doctrine Design for the Contemporary Operating Environment

As has been discussed above, Smith's 'war amongst the people' thesis is but one commentary on recent trends within the global security environment. Indeed, doctrine design reflects an array of opinions on how military ways and means might best support national strategic ends. These opinions, and in some cases orders, are generated from a variety of sources including grand strategic direction, joint doctrine, UN policy, academia and interaction with relevant non-governmental organisations (NGOs).

Grand Strategic Direction

Although land doctrine is more closely associated with joint doctrine, described below, it is heavily influenced by the direction provided within *The National Security Strategy*.[39] This sets out eight missions for Defence:

---

37 Details on the Oxford University Changing Character of War Centre and its publications and research are, available at http://www.ccw.ox.ac.uk/work-3/, accessed 1 November 2018.

38 See Theo Farrell and Terry Terriff, *The Sources of Military Change* (Boulder: Lynne Rienner, 2002), 6.

39 Her Majesty's Government, *National Security Strategy and Strategic Defence and Security Review 2015: A secure and prosperous United Kingdom* (London: Cabinet Office, 2015), available

1. Defend and contribute to the security and resilience of the UK and Overseas Territories;
2. Provide the nuclear deterrent;
3. Contribute to improved understanding of the world through strategic intelligence and the global defence network;
4. Reinforce international security and the collective capacity of our allies, partners and multilateral institutions;

Be prepared to:

5. Support humanitarian assistance and disaster response and conduct rescue missions;
6. Conduct strike operations;
7. Conduct operations to restore peace and stability;
8. Conduct major combat operations if required, including under NATO Article 5.[40]

These missions reflect the dynamic and multi-faceted global threat environment. A 'deepened and broadened' security agenda can be detected (Missions 4, 5 and 7), as can traditional concerns regarding state-based threats (Missions 1 and 8) and Cold War approaches to security, such as nuclear deterrence and collective defence (Missions 2 and 8).[41] For doctrine writers concerned with 'war amongst the people', the principal deduction is that ways and means must be configured in such a way that all missions might be executed, at times concurrently. While commentators debate which of the tasks above is the more likely, Reveron sagely observes that 'the military can no longer disaggregate war and assign certain responsibilities to particular forces or agencies. Instead, all personnel need the capability to adapt to a variety of missions'.[42] This has led to the creation of a doctrinal framework of campaign themes, operations and tactical activities. These are outlined below under the doctrinal models section.

Joint Doctrine

Joint doctrine relates to the operational level of warfare. Land tactical doctrine writers must ensure their work is compatible with joint doctrine for three reasons. First, British joint doctrine is designed to be coherent with

at https://www.gov.uk/government/uploads/system/uploads/attachment_data/file/478933/52309_Cm_9161_NSS_SD_Review_web_only.pdf, accessed 1 November 2018.

40 *National Security Strategy and Strategic Defence and Security Review 2015*, 27–29.

41 Note that, in the case of Mission 7, the Warfare Branch routinely consults with the Stabilisation Unit.

42 Derek Reveron, *Exporting Security* (Washington DC: Georgetown University Press, 2010), 70.

NATO doctrine. This ensures interoperability when conducting operations as part of the Alliance. In some thematic areas, there is no national doctrine at the joint level, with the British military deferring to NATO direction entirely. Second, alignment with joint doctrine ensures that, when land forces serve alongside members of the maritime and air components, they operate with a shared lexicon. Third, the joint level of doctrine provides an operational level gear for turning strategic direction into action at the tactical level. Consequently, joint doctrine serves as a mechanism for new ideas to enter the land tactical domain, including those relating to the themes of 'war amongst the people'.

The publication of *Joint Doctrine Publication 05 Shaping a Stable World: the Military Contribution (JDP 05)* provides an example of how thinking at the operational level on 'war amongst the people' has influenced doctrine at the land tactical level.[43] Specifically, *JDP 05* emphasised the importance of working alongside a range of actors, by means of an 'integrated' or 'full spectrum approach' (described below) when attempting to build stability overseas.[44] Linking to Smith's thesis, this approach, by using a broader range of levers than is available to the military alone, seeks to influence not only enemies but also wider audiences, actors and adversaries. The same document explores the key elements of the 'stable state', taking a sectoral approach to security, including consideration of human security.[45] Importantly, while the state remains the central referent security object within the model, non-traditional schools of thought on security are also apparent. For example, *JDP 05* identifies that 'within democratic states, human security, which requires meeting the legitimate political, economic, societal and environmental needs of individuals and groups, is equally important to protecting the state from both external and internal threats, in pursuit of national security'.[46] Alongside similar models within UN doctrine, these developments at the joint level have prompted the Warfare Branch to develop the same ideas within its *Army Field Manual Tactics for Stability Operations* and *Human Security: The Military Contribution*.

UN Policy

In 2015, *The Strategic Defence and Security Review* emphasised the UK's support for the UN and a desire to double its participation in peace support

---

43 Ministry of Defence, *Joint Doctrine Publication 05 Shaping a Stable World: the Military Contribution*, available at https://www.gov.uk/government/uploads/system/uploads/attachment_data/file/516849/20160302-Stable_world_JDP_05.pdf, accessed 1 November 2018.
44 *Joint Doctrine Publication 05 Shaping a Stable World: the Military Contribution*, 50.
45 *Joint Doctrine Publication 05 Shaping a Stable World: the Military Contribution*, 39.
46 *Joint Doctrine Publication 05 Shaping a Stable World: the Military Contribution*, 39.

operations, albeit from a low base, ostensibly due to other operational commitments.[47] Consequently, land forces have been deployed, in increasing numbers, to South Sudan and Somalia, in an effort to help keep the peace within conflicts 'amongst the people'.[48] These commitments sit alongside the more enduring commitment to peace support operations in Cyprus. To participate effectively within the UN missions, now routinely including the protection of civilians within their mandates, land doctrine has been updated.[49] Specifically, Part 3 to *Army Field Manual Tactics for Stability Operations* provides general guidance on the military contribution to peace support operations. Additionally, supported by senior leaders, certain themes of human security have been woven into land doctrine more generally.[50] These include the themes of 'Women, Peace and Security' and cultural property protection.

Reflecting UNSC Resolution 1325 and subsequent resolutions, land doctrine provides guidance on the impact of conflict on women and their role in promoting peace and stability.[51] Further, recognising that conflict affects women, men, girls and boys, the impact of violent conflict on *all* of those groups has also been highlighted. Special emphasis has been placed on the prevalence of sexual violence in conflict areas and the ways in which land forces can enable other actors to reduce it.[52] This drive is all the more important as internal research has shown soldiers, typically focused on warfighting and combat operations, are unfamiliar with these subjects. This research included questioning, by the author, of soldiers earmarked for deployments supporting the UN in South Sudan and Somalia. Such focus on human security is indicative of a willingness by the military to recognise non-traditional security agendas. This emphasis on non-combatants also aligns with Smith's view that the will of enemies and the people must be won in his new paradigm of war.

Supporting the United Nations Educational, Scientific and Cultural Organization's (UNESCO) efforts to protect cultural property in conflict,

---

47 *National Security Strategy and Strategic Defence and Security Review* 2015, 60.

48 See David Curran and Paul Williams, *The UK and UN Peace Operations: A Case for Greater Engagement*, (Oxford: Oxford Research Group, 2017), available at http://www.oxfordresearchgroup.org.uk/publications/briefing_papers_and_reports/uk_and_un_peace_operations_case_greater_engagement, accessed 1 November 2018.

49 United Nations, *Protection of Civilians Mandate* (Geneva: United Nations, 2018), available at https://peacekeeping.un.org/en/protection-of-civilians-mandate, accessed 1 November 2018.

50 Her Majesty's Government, *Continued UK Commitment to Women, Peace and Security* (London: Cabinet Office, 2015), available at https://www.gov.uk/government/news/continued-uk-commitment-to-women-peace-and-security, accessed 1 November 2018.

51 In this case, the chapter specifically refers to UNSCRs 1820, 1888, 1889, 1960, 2106 and 2122.

52 *Army Field Manual: Tactics for Stability Operations*, 10A-1.

the UK became a party to the 1954 Convention for the Protection of Cultural Property in the Event of Armed Conflict in 2017. Consequently, in addition to the provisions of LOAC, *Army Field Manual Tactics for Stability Operations* now provides detailed guidance to land forces on how to protect cultural property when operating 'amongst the people'.[53] As per the guidance provided by LOAC, the extent to which property is, or can be, protected is linked to its usage by enemy forces. Indeed, the assignation of cultural property as a security object demonstrates the challenge facing conventional armies, who have long been established to protect the state above all else. Where their focus is on the state as the principal security object, cultural security objects may not be protected, to the frustration of certain audiences, not least archaeologists.

Academia

While constrained by the direction of national security policy and the requirement to conform to joint doctrine, land doctrine writers do attempt to reconcile their work with research conducted in universities. In some cases, academics are contracted to write elements of doctrine where they hold particular expertise. In others, they are asked to comment on publications before they are circulated. Organisations routinely consulted include the Department of War Studies, King's College London, The Changing Character of War Programme at Oxford University and the Strategy and Security Institute based at Exeter University. Thus, through critical analyses of policy and doctrine, the end product can be 'improved' or at least be informed by alternative voices. For example, Griffin's 2011 critique of British stabilisation doctrine gave its doctrine writers pause when writing *Army Field Manual Tactics for Stability Operations*.[54] Equally, Sedra's critique of the conduct of Security Sector Reform (SRR) in conflict-affected states has been influential for its probing analysis of recent Western interventions in 'quasi-states', in which British land forces have participated.[55]

Non-Governmental Organisations (NGOs)

Land forces often encounter NGOs when operating 'amongst the people' in conflict zones. Consequently, a considerable effort has been

---

53 *Army Field Manual: Tactics for Stability Operations*, pp 10D.
54 See Stuart Griffin, 'Iraq, Afghanistan and the future of British Military Doctrine', *International Affairs* 87:2 (2011).
55 Mark Sedra, *Security Sector Reform in Conflict Affected Countries* (Abingdon: Routledge, 2017).

made to ensure that land tactical doctrine is informed by views from humanitarian actors to ensure cooperation is optimised. In practice, this has seen members of the Warfare Branch participate in the NGO-military contact group (NMCG) chaired by the British Red Cross. The group meets quarterly and has more than 70 members. Since 2000 it has addressed issues such as humanitarian space in Iraq, stabilisation operations in Afghanistan, civil-military coordination in Haiti and civilian protection in Syria.[56] Underpinning the NGO-military relationship are the UN's *Oslo Guidelines*, established in 1994, which seek to maintain NGO neutrality and avoid the militarisation of the delivery of aid.[57] Recent collaboration between the Warfare Branch and members of the NMCG has not only informed work on *Army Field Manual Tactics for Stability Operations* but also training for deployments to South Sudan and Somalia. Recognising the *Oslo Guidelines*, in a stability operations context, military doctrine limits land forces to providing security for humanitarian actors and, in extremis only, supporting the initial restoration of services in conflict zones.[58] Thus, land forces' ability to deliver humanitarian effect 'amongst the people' is heavily constrained.

Current Texts and Doctrinal Models

While the above section indicates influences on doctrine, this section examines doctrine itself, specifically texts and models that relate to the phenomenon of 'war amongst the people' as espoused by Smith. While they are not wholly aligned to the ideas expressed by him, some of the key themes of his thesis are perceptible.

Texts

In the land environment, *Army Doctrine Publication (ADP): Land Operations* serves as the capstone document, establishing the principles and philosophies underpinning land forces' approach to military activity. Most significantly, when describing the functions of land power, it states

---

56 Details on the British Red Cross, its activities and relations with wider government, available at http://www.redcross.org.uk/What-we-do/Protecting-people-in-conflict/Improving-civil-military-relations, accessed 1 November 2018.

57 UN Office for the Coordination of Humanitarian Affairs, *Oslo Guidelines: Guidelines on the use of foreign military and civil defence assets in disaster relief* (Oslo: United Nations Office for the Coordination of Humanitarian Affairs, 2007), available at http://www.unocha.org/sites/dms/Documents/Oslo%20Guidelines%20ENGLISH%20(November%202007).pdf, accessed 1 November 2018.

58 *Army Field Manual: Tactics for Stability Operations*, 3–3.

that 'the fundamental capability of land forces is to fight in the most demanding circumstances'.[59] This statement is vital in understanding the limits of land forces' ability to operate with sensitivity 'amongst the people'. Specifically, where a warfighting liability exists, commanders will find it challenging to prepare their soldiers for participation in operating environments requiring skills other than fighting. Of note, Lieutenant General Sir John Kiszely has observed that 'an army needs its soldiers to have a perception of themselves as something other than warriors. Without such a perception, they are liable to apply a warrior approach, for example exercising hard power when they should be exercising soft power'.[60] Despite the pre-eminence of land forces' enemy-focused fighting role, *ADP Land Operations* does articulate the ways in which they are expected to operate amongst, and protect, the people. Explicitly, the publication states that 'land forces are particularly able to secure and protect people and places persistently in the land environment. This includes providing security in support of inter-agency stabilisation and reconstruction'.[61] The detailed guidance on how land forces are expected to cover these two roles appears in lower-level doctrine outlined below.

Sitting beneath *ADP Land Operations* in the land doctrine hierarchy is 'lower-level' doctrine that describes the practices and procedures necessary for the 'effective employment of military forces'.[62] In the context of 'war amongst the people', thematic, as opposed to functional and environmental, doctrine is of most interest. Explicitly, thematic publications are divided between those concerned with warfighting and those covering stability operations. This separation has been controversial within the Army and even within the Warfare Branch itself. Indeed, the debate relates directly to how one receives Smith's thesis. Specifically, if one subscribes to Smith's paradigm shift thesis, in which he argues that war is now executed 'amongst the people', a separation of warfighting tactics from guidance on how to operate amongst people seems like folly. Conversely, if one chooses to heed Cohen and Gray's warnings regarding the potential return of massive deciding battles of interstate war, a distinct doctrine for warfighting may seem prudent. Significantly, the current doctrinal format suggests that the latter view holds sway, given that warfighting tactics are distinct from those of stability operations. Specifically, thematic doctrine is arranged as follows:

---

59   *Army Doctrine Publication Land Operations*, 1–8.
60   Lieutenant General Sir John Kiszely, 'Learning About Counterinsurgency', *Military Review* (March–April 2007), 9.
61   *Army Doctrine Publication Land Operations*, 1–8.
62   *Army Doctrine Publication Land Operations*, 1–8.

1. Army Field Manual Warfighting Tactics:
   a. Divisional Tactics
   b. Brigade Tactics
   c. Battlegroup Tactics

2. Army Field Manual Tactics for Stability Operations:
   a. Part 1. Counter Irregular Activity
   b. Part 2. The Military Contribution to Peace Support Operations
   c. Part 3. The Military Contribution to Humanitarian Assistance
   d. Part 4. The Military Contribution to Stabilisation and Reconstruction
   e. Part 5. Capacity Building.

Nonetheless, as seen later in this chapter, the complexity of the contemporary strategic and operational environments is managed, in part, through the operations themes, types of operation and tactical activity framework.

Doctrinal Models from Current Land Tactical Doctrine Pertaining to 'War Amongst the People'

Doctrine is 'what is taught'. It reflects in a single written snapshot the collective wisdom of land forces that came to the attention of its authors. Soldiers' actions are also driven by instinct, experience and what Ben-Ari has described as the use of 'folk models'.[63] The expectation is that soldiers will not follow doctrine slavishly but will apply it judiciously in the context of the environment in which they find themselves. Consequently, it is anticipated that soldiers' behaviour 'amongst the people' will be affected by multiple factors beyond the control of doctrine writers.

Operations Themes, Types of Operation and Tactical Activities

Aligned with the view expressed by Reveron – and because the *National Security Strategy* has demanded it – land forces have developed a range of ways and means to contend with the contemporary operating environment in doctrine. At the same time, following NATO doctrine, land doctrine recognises that, within a given theatre where a certain campaign theme might dominate, a range of operations are possible.[64] The campaign themes are warfighting, security, peace support and defence engagement. These themes illustrate that land forces are not focused on interstate warfare

---

63 Eyal Ben-Ari, *Mastering Soldiers* (Oxford: Berghahn, 1998), 3.
64 NATO, *Allied Joint Doctrine for the Conduct of Operations*, available at https://www.gov.uk/government/uploads/system/uploads/attachment_data/file/623172/doctrine_nato_conduct_op_ajp_3.pdf, accessed 1 November 2018.

alone. Further, within these themes, a range of types of operations might be executed. These include combat and stability operations and their sub-sets, such as counterinsurgency, peace support and humanitarian assistance. Crucially, it is possible that stability operations might occur in warfighting and combat within peace support. The salient point is that this doctrinal framework provides commanders with the flexibility to respond to the behaviour of actors, adversaries and enemies within the operating environment. In support of the varying types of operation, land forces can conduct offensive, defensive, enabling or stability activities. Again, this demonstrates that, conceptually at least, land forces have recognised the complexity of the contemporary operating environment highlighted by Smith.

Integrated Action

Another important development in land doctrine is the creation of the model of Integrated Action. In simple terms, the model informs land forces that they must consider the effect they wish to have on a given audience and the ways and means to achieve it. The same model reminds land forces that, rather than being focused on enemies alone, they may wish to have effects on other groups, such as audiences, actors and adversaries. The model also prompts land forces to consider the use of non-lethal means to influence the above groups. Consequently, it can be argued that this unifying doctrine is a clear recognition of the complexity of the operating environment described by Smith and Casey above.[65] Further, it chimes absolutely with Smith's view that it is just as important to influence civilian audiences within the operating environment as it is to exert one's will upon the enemy.

The Full Spectrum Approach

The concept of the Full Spectrum Approach emphasises that stability requires a range of actors to conduct a range of enabling activities along several lines of operation.[66] These may, for example, include political, economic and military lines. Specifically, land forces may work alongside members of the Stabilisation Unit, the Department for International Development (DfID) and host nation entities. Doctrine surrounding the Approach also describes the practices necessary to maximise land forces' interoperability with the other actors involved. The Full Spectrum Approach, though not named as such, is promoted by Smith. Referring

---

65   *Army Doctrine Publication Land Operations*, 4–3.
66   *Army Field Manual: Tactics for Stability Operations*, 2–7.

to the contemporary operating environment, he suggests that 'it must be understood the military will not be the sole or probably even lead player, and in order to achieve the best effect it is important to establish the roles of all those agencies and coordinate between them'.[67] In an academic context, it is recognised that stability is in the eye of the beholder and that Western interventions may not always be received favourably. Regardless, doctrine emphasises that land forces must be prepared for Combined, Joint, Intra-governmental, Inter-agency and Multinational (CJIIM) cooperation. This means that land forces are expected to adapt their warfighting processes to maximise interoperability when the need arises.

Conflict Sensitivity

Following land doctrine writers' consultation with the Stabilisation Unit, *Army Field Manual Tactics for Stability Operations* includes the concept of conflict sensitivity.[68] This directs land forces to consider how their intervention in a conflict-affected environment will impact upon the audiences, actors, adversaries and enemies within it. This is an important mind-set shift for military personnel who may be more accustomed to considering how they might deliver lethal effects upon their enemies. By considering the conflict dynamics more broadly, the concept of conflict sensitivity provides soldiers with additional levers of influence and enables them to participate more effectively within the Full Spectrum Approach. Consequently, in theory, it better equips them to deal with the challenges of 'war amongst the people'. Nonetheless, critics of this approach can easily draw upon the example of 'courageous restraint' as applied in Afghanistan to demonstrate its limitations.[69] Specifically, while a conflict sensitive approach might help to win 'hearts and minds' within the wider population, limiting the use of force can place land forces in significant physical danger and allow enemies to gain strength.

Factors Affecting the Use of Doctrine

Beyond doctrine writers' ability to produce guidance sufficiently nuanced for operating 'amongst the people', military audiences' subscription to it is

---

67   Smith, *The Utility of Force*, 387.

68   *Army Field Manual*, 2–13. This is derived from UK Stabilisation Unit, *Conflict Sensitivity Tools and Guidance* (London: Stabilisation Unit, 2016), available at http://sclr.stabilisationunit.gov.uk/publications/programming-guidance/1037-conflict-sensitivity-tools-and-guidance/file, accessed 1 November 2018.

69   Paul Dixon, *The British Approach to Counterinsurgency: From Malaya and Northern Ireland to Iraq and Afghanistan* (Basingstoke: Palgrave Macmillan, 2012), 136.

another matter entirely. This is influenced by a range of factors, including the presence of informal military customs and routines as well as institutional predilection for warfighting and competing lines of development.

Regarding informal military customs and routines, Ben-Ari has identified that, in addition to formal doctrine publications as a reference point, soldiers have a tendency to use experience and common sense in their daily decision making processes.[70] This is important because, when operating 'amongst the people', including other agencies involved in stabilisation, this intuition may not serve them well. Where their inclination is to warfighting, as described by Kiszely, poor choices may be made when applying the model of Integrated Action. Finally, it must be recognised that doctrine is but one line of development within land force capability. When considering 'war amongst the people', beyond doctrine, commanders must also consider matters such as training, equipment and technology, personnel, organisation and logistics. All of these take time to develop and can compete with the process of doctrinal assimilation.

Conclusion

Following interventions in Iraq, Afghanistan and Libya, PM Theresa May stated in 2017 that 'the days of Britain and America intervening in sovereign countries in an attempt to remake the world in our own image are over'.[71] Implicit in her speech was perhaps an admission that the interventions in the above states had not been successful in meeting the UK's desired ends in its security policies. This chimes with academic commentary on the 'folly' of conducting stability operations 'amongst the people' within 'quasi-states'.[72] Indeed, there have been calls, particularly in the realm of SSR, to better tailor stability operations to local conditions. In practice, this may mean less emphasis on Western Kantian notions of a liberal peace and more emphasis on reinforcing existing local governance structures. Kienscherf sees the objective of foreign state stability as being in itself problematic, noting that the Western tendency to 'problematize' the governance of conflict-affected states and to intervene with force, or the threat of it, often begets further violence.[73] These developments

---

70 Ben-Ari, *Mastering Soldiers*, 2–3.
71 Theresa May, *Speech by the Prime Minister to the Republican Party Conference on 26 January 2017* (London: Her Majesty's Stationery Office, 2017), available at https://www.gov.uk/government/speeches/prime-ministers-speech-to-the-republican-party-conference-2017, accessed 1 November 2018.
72 Sedra, *Security Sector Reform in Conflict Affected Countries*, 7.
73 Kienscherf, 'Producing 'responsible' self-governance: counterinsurgency and the violence of neo-liberal rule', 173–92.

suggest that, in the immediate future at least, stability operations involving counterinsurgency may be unattractive to British governments.

At the same time, while there is any possibility of interstate war, warfighting capabilities must be finely tuned. Regardless, while any strategic uncertainty remains, the current land doctrine set provides land forces with sufficient guidance to contend with the themes of 'war amongst the people' across the full suite of operational contexts. As ever though, land forces' success on operations depends not on slavish but judicious application of doctrine.

References

Bassford, C., 'Tip-Toe Through the Trinity or The Strange Persistence of Trinitarian Warfare', *Clauewitz.com* 14 April 2018, available at http://www.clausewitz.com/mobile/trinity8.htm#Non, accessed 1 November 2018.

Ben-Ari, E., *Mastering Soldiers* (Oxford: Berghahn, 1998).

Biden, J., *Defining the Military's Role Toward Foreign Policy: Hearing before the Committee on Foreign Relations on 31 July 2008* (Washington DC: US Senate, 2008), available at https://www.gpo.gov/fdsys/pkg/CHRG-110shrg48042/pdf/CHRG-110shrg48042.pdf, accessed 1 November 2018.

British Army, *Modernising to face an unpredictable future: Transforming The British Army* (British Army: Andover, 2012), available at https://web.archive.org/web/20130418031611/http://www.army.mod.uk/documents/general/Army2020_brochure.pdf, accessed 1 November 2018.

Buzan, B., Wæver, O., and de Wilde, J., *Security: A New Framework for Analysis* (Boulder: Lynne Rienner, 1998).

Carter, N., *Keynote speech by the Chief of the General Staff at the RUSI Land Warfare Conference on 30 June 2015* (London: RUSI, 2015), available at https://rusi.org/sites/default/files/20150630-lwc15-transcripts-cgs_opening_keynote.pdf, accessed 1 November 2018.

Carter, N., *Opening address by the Chief of the General Staff at the RUSI Land Warfare Conference on 28 June 2016* (London: RUSI, 2016), available at https://rusi.org/sites/default/files/160628-lwc16-cgs-opening_address.pdf, accessed 1 November 2018.

Casey, G. and Geren, P., *A statement on the posture of the United States Army 2008 on the 26 February 2008* (Washington DC: US Government, 2008), available at https://www.army.mil/aps/08/APS2008.pdf, accessed 1 November 2018.

Clark, C. and Kiper R., 'Strategic Thinking in an Era of Persistent Conflict', *Military Review* (May–June 2012).

Clodfelter, M., *Warfare and Armed Conflicts: A Statistical Encyclopedia of Casualty and Other Figures, 1492–2015* (Jefferson: McFarland, 2017).

Cohen, E., 'The End of War as We Know It', *The Washington Post* 21 January 2007, available at http://www.washingtonpost.com/wp-dyn/content/article/2007/01/18/AR2007011801981.html, accessed 1 November 2018.

Collins, A., 'What is Security Studies?' in Collins, A. (ed.), *Contemporary Security Studies* (Oxford: Oxford University Press, 2016).

Curran, D. and Williams, P., *The UK and UN Peace Operations: A Case for Greater Engagement* (Oxford: Oxford Research Group, May 2017), available at http://www.oxfordresearchgroup.org.uk/publications/briefing_papers_and_reports/uk_and_un_peace_operations_case_greater_engagement, accessed 1 November 2018.

Dixon, P., *The British Approach to Counterinsurgency: From Malaya and Northern Ireland to Iraq and Afghanistan* (Basingstoke: Palgrave Macmillan, 2012).

Farrell, T. and Terriff, T., *The Sources of Military Change* (Boulder: Lynne Rienner, 2002).

Ferguson, N., 'Ameliorate, Contain, Coerce, Destroy', *The New York Times* 4 February 2007, available at http://www.nytimes.com/2007/02/04/books/review/Ferguson.t.html?pagewanted=all, accessed 1 November 2018.

Fleming, C., 'New or Old Wars? Debating a Clausewitzian Future', *Journal of Strategic Studies* 32:2 (2009).

Gilmore, J., *The Cosmopolitan Military: Armed Forces and Human Security in the 21st Century* (London: Springer, 2015).

Gray, C., 'Future Warfare or the Triumph of History', *Royal United Services Institute Journal* 150:5 (2005).

Griffin, S., 'Iraq, Afghanistan and the future of British Military Doctrine', *International Affairs* 87:2 (2011).

HM Government, *National Security Strategy and Strategic Defence and Security Review 2015: A secure and prosperous United Kingdom* (London: Cabinet Office, 2015), available at https://www.gov.uk/government/uploads/system/uploads/attachment_data/file/478933/52309_Cm_9161_NSS_SD_Review_web_only.pdf, accessed 1 November 2018.

HM Government, *Continued UK Commitment to Women, Peace and Security* (London: Cabinet Office, 2015), available at https://www.gov.uk/government/news/continued-uk-commitment-to-women-peace-and-security, accessed 1 November 2018.

HM Government, *UK National Action Plan on Women, Peace and Security 2018-22* (London: HM Government, 2018), available at https://www.gov.uk/government/uploads/system/uploads/attachment_data/file/677586/FCO1215-NAP-Women-Peace-Security-ONLINE_V2.pdf, accessed 1 November 2018.

Kaldor, M., *New and Old Wars* (Oxford: Polity Press, 2001).

Kienscherf, M., 'Producing "responsible" self-governance: counterinsurgency and the violence of neo-liberal rule', *Critical Military Studies* 2:3 (2016).

Kiszely, J., 'Learning About Counterinsurgency', *Military Review* (March–April 2007).

Kitson, F., *Low Intensity Operations: Subversion, Insurgency and Peacekeeping* (London: Faber & Faber, 1971).

Lind, W., Nightengale, K., Schmitt, J., Sutton, J. and Wilson, G., 'The Changing Face of War: Into the Fourth Generation', *Marine Corps Gazette* (October 1989).

Mockaitis, T., *British Counter-Insurgency, 1919–1960* (Basingstoke: Macmillan, 1990).

Mutimer, D., 'Critical Security Studies: A Schismatic History' in A. Collins (ed.), *Contemporary Security Studies* (Oxford: Oxford University Press, 2016).

Nagl, J., *Learning to Eat Soup with a Knife: Counterinsurgency Lessons from Malaya and Vietnam* (Westport: Praeger, 2002).

NATO, *Allied Joint Doctrine for the Conduct of Operations*, available at https://www.gov.uk/government/uploads/system/uploads/attachment_data/file/623172/doctrine_nato_conduct_op_ajp_3.pdf, accessed 1 November 2018.

Peoples, C. and Vaughan-Williams, N., *Critical Security Studies* (London: Routledge, 2015).

Pfanner, T., 'Interview with General Sir Rupert Smith', *International Review of the Red Cross* 88:864 (2006).

Posen, B., *The Sources of Military Doctrine: France, Britain and Germany between the World Wars* (London: Cornell University Press, 1984).

Reveron, D., *Exporting Security* (Washington DC: Georgetown University Press, 2010).

Roberts, A., 'The Utility of Force by Rupert Smith', *The Independent* 11 November 2005, available at https://www.independent.co.uk/arts-entertainment/books/reviews/the-utility-of-force-by-rupert-smith-326177.html, accessed 1 November 2018.

Sedra, M., *Security Sector Reform in Conflict Affected Countries* (Abingdon: Routledge, 2017).

Smith, R., *The Utility of Force* (London: Allen Lane, 2005).

Strachan, H. and Scheipers, S., 'Introduction: The Changing Character of War' in Strachan, H. and Scheipers, S. (eds), *The Changing Character of War* (Oxford: Oxford University Press, 2011).

Suhrke, A., 'Human Security and the Interests of States', *Security Dialogue* 30:3 (1999).

Thompson, R., *Defeating Communist Insurgency* (London: Chatto & Windus, 1967).

UK Ministry of Defence, *Joint Doctrine Publication 05 Shaping a Stable World: the Military Contribution*, available at https://www.gov.uk/government/uploads/system/uploads/attachment_data/file/516849/20160302-Stable_world_JDP_05.pdf, accessed 1 November 2018.

UK Stabilisation Unit, *Conflict Sensitivity Tools and Guidance* (London: Stabilisation Unit, 2016) available at http://sclr.stabilisationunit.gov.uk/publications/programming-guidance/1037-conflict-sensitivity-tools-and-guidance/file, accessed 1 November 2018.

UK Stabilisation Unit, *The UK Government's Approach to Stabilisation* (London: Stabilisation Unit, 2014), available at http://sclr.stabilisationunit.gov.uk/publications/stabilisation-series/487-uk-approach-to-stabilisation-2014/file, accessed 1 November 2018.

UK Warfare Branch, *Human Security: The Military Contribution* (Warminster: Warfare Branch, 2016).

UK Warfare Branch, *Army Doctrine Publication Land Operations* (Warminster: Warfare Branch, 2016).

UK Warfare Branch, *Army Field Manual: Tactics for Stability Operations* (Warminster: Warfare Branch, 2017).

UK Warfare Branch, *Army Doctrine Publication Land Operations* (Warminster: Headquarters Field Army, 2017), available at https://www.gov.uk/government/uploads/system/uploads/attachment_data/file/605298/Army_Field_Manual__AFM__A5_Master_ADP_Interactive_Gov_Web.pdf, accessed 1 November 2018.

United Nations Office for the Coordination of Humanitarian Affairs, *Oslo Guidelines: Guidelines on the use of foreign military and civil defence assets in disaster relief* (Oslo: United Nations Office for the Coordination of Humanitarian Affairs, 2007) available at http://www.unocha.org/sites/dms/Documents/Oslo%20Guidelines%20ENGLISH%20(November%202007).pdf accessed 1 November 2018.

United Nations Trust Fund for Human Security, *Human Security* (New York: United Nations, 2018), available at https://www.un.org/humansecurity/, accessed 1 November 2018.

United Nations, *Protection of Civilians Mandate* (Geneva: United Nations, 2018), available at https://peacekeeping.un.org/en/protection-of-civilians-mandate, accessed 1 November 2018.

US Department of the Army, *Operations: Field Manual 3-0; Headquarters, Department of the Army* (Washington DC: DIANE, 2008).

Van Creveld, M., *The Transformation of War* (New York: The Free Press. 1991).

Walker, D., 'Putting "Insider-ness" to Work' in Williams, A., Jenkins, N., Rech, M. and Woodward R. (eds), *The Routledge Companion to Military Research Methods* (Abingdon: Routledge, 2016).

White, M., *Atrocities: The 100 Deadliest Episodes in Human History* (New York: Norton & Co., 2012).

CONCLUSION

Norma Rossi and Malte Riemann

> In a volume such as this discomfort is good, it generates debate and leads to change.
>
> General Sir Rupert Smith (Foreword)

This volume treats 'war amongst the people' less as a fixed and established phenomenon and more like a conceptual prism through which contemporary intra-state conflicts can be read and questioned. As such, it considers it not as a monolithic 'new type' of war, but as a framework that can shed light on complex networks and dynamics and their context dependency. Taken together, the different contributions therefore provide not only new empirical insights on 'war amongst the people' but, through this prism, encourage novel ways of seeing and assessing it. It is the aim of this conclusion to capture some of the implications that arise from the contributions across the four analytical lenses identified in the introduction – the conceptual, the practical, the legal and the domestic – and show how the contributions create shared themes across the four dimensions, thereby recomposing the prism. A key tenet that the prism illuminates is how politics and war overlap in 'war amongst the people', which will be considered in more depth in the final part of this conclusion.

Before examining the shared implications of the four dimensions, it is apposite to summarise briefly the key points of each dimension. In the first three chapters, Heuser, Raitasalo and Waterman engaged with the 'war amongst the people' paradigm on a conceptual level. In her considerations on the phenomenon called 'war', Heuser stresses that this phenomenon is highly culture-dependent and therefore each war's peculiarities and distinctive circumstances need to be fully appreciated. In broad agreement with Heuser's claim that war is culture-dependent,

Raitasalo's contribution shows how Western states have reconceptualised 'war' and international security in the post-Cold War era, which leads to an increased focus on engaging in intra-state conflicts. This, he argues, not only made Western states neglect the traditional warfighting and deterrence capabilities of their armed forces, but also embroiled them in conflicts that are of little or no connection to their strategic security interests. In line with Heuser and Raitasalo, Waterman emphasises the complexity of war and how the understanding of this phenomenon displays difference over space and time. He develops this claim through a reading of how 'the people' are understood in 'war amongst the people', arguing that this has often been over-simplistic, thereby ignoring the complex relations, conflicts and variety of actors that make up the 'people'.

The next three chapters depart from primarily conceptual analysis, focusing instead on the practical challenges emanating from contemporary conflicts. Specific consideration was given to the actors fighting such wars. Grespin and Holmes do so by dealing with the central question of how to train and educate the military to face the complexities of 'war amongst the people'. In their analyses, both acknowledge the insufficiencies of tactical training if not combined with a more comprehensive approach that promotes strategic change for long-term sustainability and transformation. Moving away from regular forces, Rauta's contribution considers the irregular elements in 'war amongst the people'. He conceptualises irregular forces through the distinction between proxies and auxiliaries, demonstrating how a more multi-layered approach to understanding irregular forces is essential for developing a more effective strategy to fight in irregular conflicts.

Questions concerning the conceptualisation of actors in 'war amongst the people' also lie at the heart of Melancon's and Davies' contributions on the legal dimensions. Originally designed for conventional interstate war, LOAC has been challenged to adapt to the growing intra-state nature of contemporary conflicts. Here, 'war amongst the people' has raised a series of pressing legal questions. Two aspects are particularly salient: first, 'war amongst the people' reveals a central legal dualism between those people who take part in conflict and those who do not. Ensuring this distinction is central not only to the legality of the conflict, but also to its legitimacy and political sustainability. Secondly, the rights of detainees raise key questions regarding the relationship between fighting 'wars amongst the people' and respecting human rights. This speaks directly to the problem of which rights should apply, namely those specific to war or those pertaining to peace, as identified by Heuser in the first chapter.

In the final two chapters, Wilson and Bailey turn their attention to the 'intervening' state rather than the 'intervened' – thereby shedding light on an often overlooked, yet significant, aspect of 'war amongst the people' – the domestic context of the intervening state. To elucidate their claims, they concentrate on the UK as a case study. In his chapter, Wilson examines variations in popular consent for different 'wars amongst the people' (Afghanistan, Iraq, Libya and, as appropriate, Syria). Bailey adopts a different perspective, investigating the impact of the 'war amongst the people' paradigm on British Army doctrine. Although both contributions focus on very different topics, they illustrate and reiterate the importance of linking the objectives set by civilian authorities with the development of military strategy (Wilson) and military doctrine (Bailey).

Common Themes and Questions

Four common themes, and the questions they pose, not just in relation to the 'war amongst the people' paradigm, but in thinking about war and military engagements more generally, emerge as a result of the analysis contained within this volume. The first theme relates to the concept of 'the people' and how this should be understood. How are they created, shaped and re-shaped by war? This resonates with the second theme: the question of knowledge. How can conflicts categorised as 'war amongst the people' be understood and analysed when these display an absence of constants and binaries, on which legal and military thinking has traditionally relied? For example, 'war amongst the people' defies clear distinctions between dichotomies such as peace/war and combatant/non-combatant. The continued absence of these constants and binaries – despite the best efforts of the literature to revise and reconsider the concepts, as with this volume – points to issues related to the third theme: the difficulties in linking the tactical with the strategic object in 'war amongst the people'. If 'wars amongst the people' – as Smith reminds us – are 'sub-strategic', how can this strategic gap be overcome? How can a clear strategic policy be formulated if 'war amongst the people' places the emphasis on the tactical over the strategic? Furthermore, how can success in the absence of strategy be achieved or, indeed, even be identified? The latter builds the framework for the last theme: the question of local and international legitimacy. The question of achieving and maintaining local and international legitimacy is a spectre that haunts this volume's different contributions. What provides the veneer for legitimacy in 'war amongst the people'? How can it be achieved and what are the constraining factors? The remainder of this chapter will unpack each of these themes in separate sections. The final section will then elaborate on how 'war amongst the people' affects the relationship between war and politics, showing how these specific wars both disorder and unsettle politics and people, while simultaneously also transforming and generating them.

People Are Not Found, They Are Made

Following Barkawi and Brighton, '[w]hile destructive, war is a generative force like no other. It is of fundamental significance for politics, society and culture'.[1] As such, war is not only influenced by the social, economic, political and cultural context in which it takes place, but it also re-shapes, reproduces and changes these contexts. War therefore has a key function in actively generating and constituting the social realm and the people within it.[2] This is not only valid for the space in which war physically takes place, but also for the societies, states and groups that intervene and potentially become enmeshed in such conflicts. While diverse in their outlook, the contributions to this volume demonstrate war's generative power by showing how 'wars amongst the people' are constituting the social and political realities of war. Indeed, as Heuser's contribution shows, the very definition of the existence of a state of war is produced by the actors involved in it: 'In the end it is a subjective and political decision and not an apolitical and objective evaluation of factors that determines whether one is or considers oneself to be "at war"'. Therefore, the knowledge about war is not independent or detached from the very act of war, as 'knowledge about war is never fully exterior to an order war itself creates'.[3] And here, like war, the people are never a neutral or static object, but always the product of distinct contexts, knowledges and relations. Waterman's contribution makes this point clear. In Chapter 3 he shows how 'war amongst the people' is not only the site of a war between and amongst pre-existing people(s), but also the very site of creation of such peoples. In other words, as most prominently pointed out by the 'new wars' school, the identity of separate peoples does not simply pre-date conflicts, but it is created by identity politics that are deployed during the conflict as a mobilising strategy.[4] A key implication of this, according to Waterman, is the need to 'connect theory to the politics of insurgency and COIN, exploring the processes

---

1   Tarak Barkawi and Shane Brighton, 'Powers of War: Fighting, Knowledge and Critique', *International Political Sociology* 5:2 (2011), 126.

2   Barkawi and Brighton, 'Powers of War', 126–43.

3   Barkawi and Brighton, 'Powers of War', 135.

4   Mary Kaldor, *New and Old Wars: Organized Violence in a Global Era* (Cambridge: Polity Press, 1999).

of contestation, coercion, accommodation and bargaining that take place between counterinsurgents and these different socio-political groupings'.

The productive function of 'war amongst the people' emerges most clearly from Holmes' contribution. Drawing upon Bourdieu's work, Holmes exposes how training and education to conduct 'war amongst the people' is also a terrain of creation, negotiation and contestation of the values and norms that govern it. Referring to the productive power of the pre-deployment training space, Holmes argues that, not only is this space 'a militarised field of power, but it is also a field of education wherein identities and roles (or subject positions) are constructed and negotiated during transfers of knowledge between educators and learners'. Holmes' analysis of the training space shows that female peacekeepers are not passive receivers of UN training, but co-participants in the creation and interpretation of the normative framework, which they are asked to learn. This highlights an important aspect of 'war amongst the people', namely that the creative powers of war are not limited to affecting local peoples, but also form, transform and shape the external intervening forces. Indeed, as Wilson reminds us, 'war amongst the people' is not only productive in 'creating' the people in theatre, but also creates the people 'at home'. Given the range of effects that 'war amongst the people' can have on the domestic society of intervening states, these are critical when it comes to shaping political consent and people's attitudes towards external interventions and conflict. In this, Wilson confirms Michael Dillon's observations with regards to the 'Liberal Way of War': 'War forms and transforms governmental institutions and practices as it does political rationalities and civic cultures'.[5] By treating 'war amongst the people' as a prism for analysis, the contributions to this book expose how the very people that it claims to be conducted amongst are not *found* but rather *made* and reveals these conflicts to be fields of complex dynamics and practices that produce the social reality in which they are conducted.

Epistemology, Complexity and the Absence of Constants

That 'wars amongst the people' actively construct the social reality in which they take place leads to a further theme that connects the contributions to this volume: the *status* of 'knowing' and 'understanding'. In their different approaches to investigating 'war amongst the people', the chapters reveal

---

5   Michael J. Dillon, 'Introduction: From Liberal Conscience to Liberal Rule' in Michael J. Dillon and Julian Reid (eds), *The Liberal Way of War: Killing to Make Life Live* (London: Routledge, 2009), 9.

a strong tension between, on the one hand, the exigencies of knowing and understanding 'war amongst the people' and, on the other, that of practically finding ways to engage in them. Regarding the former, Heuser argues that, to understand 'war amongst the people', there is a need to break with the dualisms and binaries that dominate conventional understandings of conflict and, instead, embrace the culturally, contextually and historically ever-changing realities of war. In line with this, Waterman claims that easy categorisations are untenable. Studying the people(s) requires paying attention to the diverse, and, at times, opposing motivations, interests and identities that shape them. In the move from 'war amongst the people' to 'war amongst the peoples', he proposes an approach that defies the reification of the people in a uniformed and undifferentiated mass and instead requires embracing the complexity and context-specific character of each conflict and the people(s) involved.

Yet, sharp categorisations and binary oppositions are essential in military thinking and planning, a paradox that remains difficult to resolve. The most obvious example of this is the opposition drawn between 'self' and the 'military other', most commonly categorised as 'the enemy'. As Christensen et al. argue, within 'war amongst the people':

> similar oppositions are constantly being drawn, for instance between 'friendly forces' and 'hostile forces', and between civil and military domains. When faced with an enemy who blends and blurs with the population, such oppositions limit the understanding of the enemy as well of the local inhabitants, and thereby limit the potential for capturing the enemy and of winning the 'hearts and minds' of the population.[6]

Melancon and Rauta's chapters draw attention to this tension. While sharing with Heuser and Waterman the need to acknowledge the complexities of 'war amongst the people' and the fallacies of binary categorisations, they also argue for the necessity of finding ways to (re) draw lines that allow nonetheless for clear distinctions in order to engage in 'war amongst the people' on a more practical level. Indeed, in her chapter, Melancon re-constitutes a firmer binary distinction between combatant and non-combatant in order to obtain a more applicable definition of 'Direct Participation in Hostilities'. In doing so, she encounters the first paradox outlined in Heuser's chapter, namely that 'much thinking and writing about war revolves around attempts to impose rules on what is the extreme of unruly behaviour'. Guided by a similar concern, Rauta tries to

---

6   Maya Christensen, Rikke Haugegaard and Poul Martin Linnet, *'War amongst the people' and the absent enemy: Towards a cultural paradigm shift* (Copenhagen: Royal Danish Defence College Publishing House, 2014).

create distinct analytical definitions to recognise and distinguish between proxies and auxiliaries. He defines proxies as actors with autonomous strategic and political aims and auxiliaries as actors that display only a tactical-military outlook on the conflict. Although Melancon and Rauta focus on very different aspects of 'war amongst the people', both share the concern that the practical demands of fighting 'wars amongst the people' necessitate binary distinctions, such as those between combatants/non-combatants and proxies/auxiliaries. This is in opposition to Waterman's and Heuser's contributions, which argue that reinstating binaries and dualities is unhelpful in understanding 'war amongst the people'. While in disagreement about the value of binary distinctions per se, all four authors accept that such a debate points at the paradoxical relationship between theory and practice, which makes engaging in 'war amongst the people' so challenging.

The Link Between Strategy and Tactics

Different chapters of this volume show that Smith's central concern for linking the tactical with the strategic is still an unresolved problematique in contemporary 'war amongst the people'. As he remarked, 'conflicts are sub-strategic; frequently only tactical, in effect'.[7] Crucially, as Raitasalo, Rauta, Holmes and Grespin all show, the link between the tactical and the strategic is an epistemological as well as a practical necessity. In other words, while it is impossible to understand or fight a 'war amongst the people' without integrating strategic thought into tactical decision-making, there still seems to be a strategic gap in the ways in which 'war amongst the people' is understood. Raitasalo makes this point clearly in his analysis of the centrality of strategy – and lack thereof – in the decision-making process of Western states engaging in such conflicts. Holmes too highlights the limited effects of UN training programmes, which focus solely on the tactical level and lack an engagement with the strategic level. As she shows, tactical training offers quick fixes, but it does not correspond to the long-term and transformative aims that are required for sustainable success. For her part, Grespin warns against the temptation of finding quick and cheap fixes, by prioritising tactical-military training over a more strategic and holistic approach to nation-building, which she sees happening in the US government's decision to significantly privatise such

---

7   Rupert Smith, 'Thinking about the Utility of Force in War Amongst the People' in John Andreas Olsen (ed.) *On New Wars* (Oslo: Norwegian Institute for Defence Studies, Institutt for Forsvarsstudier2007), 33.

training efforts. Similarly, both Wilson and Bailey highlight the criticality of filling this strategic gap, to maintain public support and, even more importantly, to prepare armed forces for engagement in 'war amongst the people'. In particular, Bailey shows this through the British Army's attempts to adapt its doctrine to the changing strategic context shaped by 'war amongst the people', amongst other factors. His argument is in line with what Hutchings and Treverton have observed with regards to the implementation of strategy: '[s]trategy needs to start at the top', but 'any strategy is implemented by hundreds or thousands of lower-ranking officials'.[8] This highlights the continued and important link between the tactical and the strategic.

While these contributions seem to agree on the necessity of filling the strategic gap in 'war amongst the people', they also reveal a point of potential disagreement: how should 'our' strategic aims be defined? Raitasalo, on the one hand, and Grespin and Holmes on the other, seem to point towards two different answers. Grespin sees nation-building as a central pillar of the long-term strategic approach to 'war amongst the people'. Holmes similarly highlights the importance of long-term approaches, focusing on norm transformation in relation to gender inclusion and gender equality, which she considers to be a key component in the overall strategic approach to such interventions. Both, therefore, put liberal democratic values at the centre of defining strategic goals. This is in line with what Rob Johnson has recently argued should be an integral part of Western strategic approaches to contemporary conflicts:

> the exercise of free speech; the refusal to accept that any person, in whatever high office, can be above the law; the protection of the citizen from arbitrary imprisonment; the right to be represented through parliamentary democracy; equality in opportunity; and the freedom of conscience, to name just a few of our most fundamental principles ... these ideas have made liberal democracy a success.[9]

Conversely, Raitasalo prioritises a strategic approach grounded in geopolitics and Realpolitik, in which the strategic outlook should adhere to the limited and clearly demarcated national interests of intervening state(s). His argument goes even further, questioning whether there should

---

8   Robert L. Hutchings and Gregory F. Treverton, *Rebuilding Strategic Thinking: A Report of the CSIS Transnational Threats Project* (Washington DC: Center for Strategic and International Studies, October 2018), 30, available at https://espas.secure.europarl.europa.eu/orbis/sites/default/files/generated/document/en/181018_RebuildingStrategicThinking_WEB_v2.pdf, accessed 29 December 2018.

9   Rob Johnson, 'The Changing Character of War', *The RUSI Journal* 162:1 (2017), 8.

even be a focus on 'war amongst the people', both doctrinally and in terms of the allocation of intellectual and material resources, particularly if this comes at the expense of preparing for conventional warfare. Indeed, in a recent contribution to the *National Interest,* he argues that, through the focus on 'counterterrorism, irregular warfare, counterinsurgency operations and military crisis management … much of the know-how and military ethos related to defense and deterrence have been lost'.[10] While a refocusing on conventional warfighting capabilities seems to resonate with strategy-makers in a climate of tensions between the US, China and Russia, Mary Kaldor warns that the failure to deal with intra-state wars could increase the likelihood of interstate war: 'Of course, a return to old wars cannot be ruled out … But failure to deal with the "new wars" of the present might make that possibility more plausible'.[11] The initially localised conflict in Syria that, as it progressed, drew both the US and Russia in, is a case in point. Syria therefore highlights that a perspective emphasising an either/or choice between a strategic outlook focusing on conventional warfare or 'war amongst the people' ultimately poses a false choice, as it ignores the fact that the two can be and regularly are highly interconnected. This important debate is destined to continue, as disagreement concerning what should constitute strategic aims forms a significant aspect of a much wider issue, given that the constitution of strategic aims ultimately defines success.[12] Indeed, recent campaigns have shown that the very meaning of success has frequently been unclear;[13] at times too narrowly focused on tactical encounters, at others too broad in scope, exemplified most vividly by the mantra of 'democracy promotion'. This has direct consequences for the very possibilities of engaging in 'war amongst the people', as Wilson's chapter reminds us. 'Success sells' in the sense that it is a vital element in sustaining domestic support for military interventions.

Local and International Legitimacy

Directly linked to the problem of success in 'war amongst the people' is the problem of local legitimacy. While the 'hearts and minds'

---

10 Jyri Raitasalo, 'Big War is Back', *The National Interest* (September 2018), available at https://nationalinterest.org/feature/big-war-back-30802, accessed 29 December 2018.

11 Mary Kaldor, 'In Defence of New Wars', *Stability: International Journal of Security and Development* 2:1 (2013), 5.

12 Colin S. Gray, *Modern Strategy* (Oxford: Oxford University Press, 1999).

13 Colin S. Gray, *Defining and Achieving Decisive Victory* (Carlisle, PA: Strategic Studies Institute, 2014); William C. Martel, 'Victory in scholarship on strategy and war', *Cambridge Review of International Affairs* 24.3 (2011), 513–36.

catchphrase is repeatedly evoked, the literature on small wars has begun to turn its attention to wider complexities concerning legitimacy. This recalibration has grown out of the dispelled notion that elections alone can build legitimacy. The facts on the ground in recent counterinsurgency operations have painfully demonstrated that a sole focus on elections, what Greene has labelled the 'pathology' of COIN, is not enough.[14] The contributions to this volume raise some critical points regarding this matter. Heuser's call to abandon 'the crass idea that peoples and cultures are interchangeable' and therefore to acknowledge the culturally context-specific nature of 'wars amongst the people' directly connects with the focus of recent counterinsurgency literature on the locally and culturally bounded nature of legitimacy.[15] Additionally, the legal contributions to this volume point to the centrality of legality in building legitimacy within 'wars amongst the people'. Melancon's chapter shows that Direct Participation in Hostilities should be re-drawn to allow for a more nuanced and permissive targeting of specific actors in an attempt to match the current LOAC framework with the demands dictated by state practices. Here, according to Melancon, the problem lies not with the lack of international legislation on defining DPH, but with intervening states' perception that this is not functional in fighting within 'wars amongst the people'. This negatively affects the legitimacy of these legal frameworks. For the purpose of building local and international legitimacy, Melancon argues that it is important to modify this framework in ways that still limit collateral damage, while simultaneously being perceived as effective and applicable by the intervening state. Davies approaches the issue from a different standpoint, claiming that international legislation governing detention in NIAC is scarce: 'The increase in the number of NIACs is inversely proportionate to the quantity of substantive international humanitarian law rules governing them'. In his analysis of which legislation is best suited for detainment practices in 'war amongst the people', Davies sheds light on another central issue concerning the problem of local and international legitimacy, namely the relationship between victory and the respect for international human rights. This raises the question of whether it is necessary to strike the right balance between both or whether these objectives are intrinsically linked and therefore mutually dependent. The answer to this question is connected to what the external actor ultimately wants to achieve: power or authority

---

14   Samuel R. Greene, 'Pathological Counterinsurgency: the failure of imposing legitimacy in El Salvador, Afghanistan, and Iraq', *Third World Quarterly* 38:3 (2017), 563–79.

15   Andrew J. Gawthorpe, 'All Counterinsurgency is Local: Counterinsurgency and Rebel Legitimacy', *Small Wars and Insurgencies* 28:4/5 (2017), 839–52.

over the people in 'war amongst the people', as '[a]uthority ... is a form of power – but power resting on recognition based on legitimacy, rather than on coercion or material incentives'.[16] Evidence from recent studies on counterinsurgency strongly point towards the advantages of authority over power. In the words of Gawthorpe, 'counterinsurgency involves not just an attempt to bolster the legitimacy of an incumbent state, but ... it is fundamentally a struggle between competing legitimacies'.[17] This is linked to Smith's observation that 'wars amongst the people' are contests over legitimacy in which the overall political objective of the conflicting parties is to win 'the will of the people'.

Yet, capturing 'the will of the people' is not without its difficulties. First, the quest for local legitimacy forces an army to confront an extreme version of Heuser's second paradox of war: 'that during war, they may be called upon at different stages or in very short succession to fulfil utterly contradictory duties, from the infliction of extreme violence to extreme restraint'. This tension is also evident in Bailey's examination of recent conflicts and their impact on doctrine. The doctrinal changes these lead to have often required armed forces to perform multiple and, at times, even contrasting functions. Second, attempts to find shortcuts to authority-building can have enormous limitations, as Grespin warns with regards to the focus of intervening states on the delivery of tactical training at the expense of a more holistic approach to state- and nation-building. Third, Wilson's analysis shows that the domestic context of the intervening state works like a quickly emptying hourglass in which the length of the conflict negatively affects public support. This makes long-term authority-building extremely difficult, but it is nevertheless extremely important with regards to both local and international legitimacy.

Politics and War in the 'War Amongst the People'

Taken together, the chapters in this volume constitute a professional practitioner and scholarly reflection on four facets that emanate from modern conflicts fought within and amongst populations: the conceptual, the practical, the legal and the domestic. In this, they offer novel perspectives on timely and important concerns that emerge from engaging with 'war amongst the people' and expose four common themes and several critical questions.

---

16   Dominik Zaum, 'International Transitional Administrations and the Politics of Authority Building', *Journal of Intervention and State Building* 11:4 (2017), 410.

17   Gawthorpe, 'All Counterinsurgency is Local', 840.

Questions concerning the identity of actors, the absence of constants, the link between the tactical and the strategic, as well as the quest for local and international legitimacy, all revolve around the relationship between politics and war. A reflection on this aspect is of pressing concern, as recent 'wars amongst the people' have painfully demonstrated that victory in battle has all too often not been translated into an overall strategic and political victory. Indeed, as Smith has suggested in his foreword to this volume, a key characteristic of 'war amongst the people' is precisely the paradox that tactical victories on the battlefield do not necessarily correspond with strategic success at the political level. This is why '[y]ou can win every fight and lose the war'.[18]

The reason for this is found in the very character of 'war amongst the people', insofar as these do not respect the spatial and temporal configurations that regulate strategy-making in war. Following Andrew Carr, the very possibility for strategy depends on clear and well defined temporal and spatial dimensions, as '[s]trategy is action in space and time'.[19] First, from a spatial dimension, strategy requires a clear distinction between the space of peace in which politics dominate and a space of war in which normal politics and law are suspended in the name of the exceptional laws of war and military action.[20] Second, the temporal dimension requires the existence of a clear distinction between times of war and times of peace. In its most traditional form, these times were sanctioned respectively by declarations of war and the signing of peace treaties that terminated hostilities. Yet, 'wars amongst the people' do not respect these spatial and temporal configurations. Rather than displaying a clear separation between times of war and times of peace, 'war amongst the people' tends to be 'persistent and protracted'[21] and the spatial boundaries between peace and war become blurred.[22] Indeed, the spaces in which 'war amongst the people' takes place become affected by chronic insecurity and violence.[23]

---

18 Rupert Smith, 'Methods of Warfare', *International Review of the Red Cross* 88:864 (2006), 719–27.
19 Andrew Carr, 'It's about time: Strategy and temporal phenomena', *Journal of Strategic Studies* 41:6 (2018).
20 See Carl Schmitt, *Political Theology: Four Chapters on the Concept of Sovereignty* (Chicago: University of Chicago Press, 2004); Giorgio Agamben, *State of Exception* (Chicago: University of Chicago Press, 2005).
21 Mary Kaldor, *The New Peace: A lecture by Mary Kaldor* (Somerville: World Peace Foundation, November 2012), available at https://sites.tufts.edu/reinventingpeace/2012/11/28/the-new-peace-a-lecture-by-mary-kaldor/, accessed 29 December 2018.
22 See Development, Concepts and Doctrine Centre, *Global Strategic Trends: The future starts today* (Shrivenham: Development, Concepts and Doctrine Centre, 2018), 125, available at https://www.gov.uk/government/publications/global-strategic-trends, accessed 29 December 2018.
23 Kaldor, *The New Peace*.

The effect of this spatial and temporal fragmentation ultimately upsets the Clausewitzian relationship between politics and war. Following Clausewitz, war is the continuation of politics by other means, meaning that one (war) is functional to the other (politics).[24] This implies that war is a function of politics and requires that the relationship between the two is regulated by a temporal sequencing: 1. politics – 2. war as a continuation of politics by other means – 3. peace and then the cycle (potentially) begins anew. Yet 'war amongst the people' breaks both the temporal and spatial sequencing of war. The effect leads to social fragmentation, as Gawthorpe has observed: 'The segmentation of physical space which accompanies irregular warfare is accompanied by a parallel segmentation of social, political, and economic space. This also leads to a segmentation of legitimacy dynamics.'[25] This is most clearly seen in the conflation of military and civilian spheres and the dissolving boundaries between combatants and non-combatants in 'war amongst the people'. Without a clear separation between civilian and military spheres, nor a clear temporal separation between times of peace and times of war, the patterns of life and death overlap. As such, rather than one being the function of the other, war and politics overlap in the same space and most if not all of the time. In the Clausewitzian paradigm, a political decision leads to military action (hence a temporary suspension of politics), which subsequently has a political effect (i.e. a peace treaty). In 'war amongst the people', this sequencing is interrupted and military actions are already politics rather than the suspension of it. In other words, military operations do not have subsequent political and strategic effects, but, instead, these become, *from the very beginning*, political actions, which have immediate effects on the societies in which they are enacted. As Amanda Hester summarises Smith's argument, 'every action performed by every soldier, at every level, holds strategic significance'.[26] Through this process the armed forces are placed at the centre of the political arena. Ultimately, this might lead to an inversion of the relationship between war and politics, in which war is no longer the continuation of politics by other means, but politics becomes the continuation of war by other means.[27]

While the common themes and questions this volume exposed can be daunting, they are also what makes the continued examination of the 'war

---

24 Carl von Clausewitz, *On War* (Oxford: Oxford University Press, 2007).
25 Gawthorpe, 'All Counterinsurgency is Local', 843.
26 Amanda Hester, *Beyond an Enemy: Exploring the need for mindfulness training in a new generation of warfare* (Nova Scotia: Royal United Services Institute of Nova Scotia, November 2017), available at https://rusi-ns.ca/wp-content/uploads/2017/12/Beyond_an_Enemy.pdf, accessed 29 December 2018.
27 Michel Foucault, *"Society Must be Defended": Lectures at the Collège De France 1975–1976* (London: Picador, 2003), 15.

amongst the people' paradigm a relevant and pressing issue. If read through the prism of the four lenses, the findings of this volume have profound implications at both the epistemological and practical level. First, they point to the constitutive character of these wars, both for those living inside the spaces of conflict and those intervening from the outside. Second, they reveal an essential relationship (albeit one filled with tensions) between the theory and practice of 'war amongst the people'. Third, they show the centrality of working towards bridging the strategic gap in 'war amongst the people', while simultaneously exposing contrasting ways of defining strategic success. Finally, they link the possibility for strategic success to the practical need to create local legitimacy, exposing it as an area in need of further investigation.

As Rupert Smith suggests in his foreword, a volume like this should leave those in charge of writing UK strategy and British military doctrine with a sense of discomfort, for, in spite of strong attempts to re-think the role and scope of the British Army in contemporary conflicts, the task requires an on-going critical re-evaluation and elaboration. This volume provides a useful and timely contribution to this task by creating a productive dialogue between a range of different and interested parties.

References

Agamben, G., *State of Exception* (Chicago: University of Chicago Press, 2005).

Barkawi, T. and Brighton, S., 'Powers of War: Fighting, Knowledge and Critique', *International Political Sociology* 5:2 (2011).

Carr, A., 'It's about time: Strategy and temporal phenomena', *Journal of Strategic Studies* 41:6 (2018).

Christensen, M., Haugegaard, R. and Martin Linnet, P., *'War amongst the people' and the absent enemy: Towards a cultural paradigm shift* (Copenhagen: Royal Danish Defence College Publishing House: 2014).

von Clausewitz, C., *On War* (Oxford: Oxford University Press, 2007).

Development, Concepts and Doctrine Centre, *Global Strategic Trends: The future starts today* (Shrivenham: Development, Concepts and Doctrine Centre, 2018).

Dillon, M.J. and Reid, J. (eds), *The Liberal Way of War: Killing to Make Life Live* (London: Routledge, 2009).

Foucault, M., *"Society Must be Defended": Lectures at the Collège De France 1975–1976* (London: Picador, 2003).

Gawthorpe, A.J., 'All Counterinsurgency is Local: Counterinsurgency and Rebel Legitimacy', *Small Wars and Insurgencies* 28:4/5 (2017).

Gray, C.S., *Modern Strategy* (Oxford: Oxford University Press, 1999).

Gray, C.S., *Defining and Achieving Decisive Victory* (Carlisle: Strategic Studies Institute, 2014).

Greene, S.R., 'Pathological Counterinsurgency: the failure of imposing legitimacy in El Salvador, Afghanistan, and Iraq', *Third World Quarterly* 38:3 (2017).

Hester, A., *Beyond an Enemy: Exploring the need for mindfulness training in a new generation of warfare* (Nova Scotia: Royal United Services Institute of Nova Scotia, November 2017), available at https://rusi-ns.ca/wp-content/uploads/2017/12/Beyond_an_Enemy.pdf, accessed 29 December 2018.

Hutchings R.L. and Treverton, G.F., *Rebuilding Strategic Thinking: A Report of the CSIS Transnational Threats Project* (Washington DC: Center for Strategic and International Studies, October 2018), available at https://espas.secure.europarl.europa.eu/orbis/sites/default/files/generated/document/en/181018_RebuildingStrategicThinking_WEB_v2.pdf, accessed 29 December 2018.

Johnson, R., 'The Changing Character of War', *The RUSI Journal* 162:1 (2017).

Kaldor, M., *New and Old Wars: Organized Violence in a Global Era* (Cambridge: Polity Press, 1999).

Kaldor, M., 'In Defence of New Wars', *Stability: International Journal of Security and Development* 2:1 (2013).

Martel, W.C., 'Victory in scholarship on strategy and war', *Cambridge Review of International Affairs* 24:3 (2011).

Olsen, J.A. (ed.), *On New Wars* (Oslo: Norwegian Institute for Defence Studies, Institutt for Forsvarsstudier, 2007).

Raitasalo, J., 'Big War is Back', *The National Interest* (September 2018).

Schmitt, C., *Political Theology: Four Chapters on the Concept of Sovereignty* (Chicago: University of Chicago Press, 2004).

Smith, R., 'Methods of Warfare', *International Review of the Red Cross* 88:864 (2006).

Zaum, D., 'International Transitional Administrations and the Politics of Authority Building', *Journal of Intervention and State Building* 11:4 (2017).

INDEX

NOTES

www.ingramcontent.com/pod-product-compliance
Lightning Source LLC
Chambersburg PA
CBHW060313030426
42336CB00011B/1018